# BOOTS ON THE GROUND

## Disaster Response in Canada

*Johanu Botha*

Over the last century, the scale of Canada's domestic disaster response system has grown significantly due to the country's increased capacity for emergency management and the rise in natural hazards. However, there has been no systematic assessment of how effectively this multilevel system, which includes all levels of government and the military, has been integrated, and how efficient this system actually is at responding to high-level disasters.

Using in-depth archival analysis and interviews with senior military and civilian officials on the inside, *Boots on the Ground* provides a detailed examination of Canada's disaster response system. Including policy recommendations focused on the expansion of emergency management networks, the maintenance of Canada's decentralized emergency management system, and disaster response resources for First Nations communities, *Boots on the Ground* aims to highlight opportunities to improve Canada's urgent disaster response.

*Boots on the Ground* offers helpful lessons for students, policy makers, emergency management practitioners, and military officers, ensuring that readers gain concrete insights into the strategic and efficient implementation of disaster response initiatives.

JOHANU BOTHA leads the Manitoba Emergency Measures Organization as Assistant Deputy Minister with the Government of Manitoba.

JOHANU BOTHA

# Boots on the Ground

## Disaster Response in Canada

UNIVERSITY OF TORONTO PRESS
Toronto Buffalo London

© University of Toronto Press 2022
Toronto Buffalo London
utorontopress.com

ISBN 978-1-4875-2977-2 (cloth)   ISBN 978-1-4875-2980-2 (EPUB)
ISBN 978-1-4875-2978-9 (paper)   ISBN 978-1-4875-2979-6 (PDF)

**Library and Archives Canada Cataloguing in Publication**

Title: Boots on the ground : disaster response in Canada / Johanu Botha.
Names: Botha, Johanu, author.
Identifiers: Canadiana (print) 2022013314X | Canadiana (ebook) 20220133182 | ISBN 9781487529772 (hardcover) | ISBN 9781487529789 (softcover) | ISBN 9781487529802 (EPUB) | ISBN 9781487529796 (PDF)
Subjects: LCSH: Emergency management – Canada. | LCSH: Disaster relief – Canada.
Classification: LCC HV551.5.C3 B68 2022 | DDC 363.340971–dc23

We wish to acknowledge the land on which the University of Toronto Press operates. This land is the traditional territory of the Wendat, the Anishnaabeg, the Haudenosaunee, the Métis, and the Mississaugas of the Credit First Nation.

This book has been published with the help of a grant from the Federation for the Humanities and Social Sciences, through the Awards to Scholarly Publications Program, using funds provided by the Social Sciences and Humanities Research Council of Canada.

University of Toronto Press acknowledges the financial support of the Government of Canada, the Canada Council for the Arts, and the Ontario Arts Council, an agency of the Government of Ontario, for its publishing activities.

Canada Council   Conseil des Arts
for the Arts     du Canada

ONTARIO ARTS COUNCIL
CONSEIL DES ARTS DE L'ONTARIO
an Ontario government agency
un organisme du gouvernement de l'Ontario

Funded by the    Financé par le
Government      gouvernement
of Canada       du Canada

Canadä

# Contents

*List of Tables* vii

*Acknowledgments* ix

Introduction 3

1 Emergency Management and the Military in Canada 10
2 Assessing Disaster Response through an Original Collaborative Framework 35
3 The *Presence* of Interorganizational Collaboration 62
4 The *Quality* of Interorganizational Collaboration 87
5 The *Barriers* to Interorganizational Collaboration 110
6 Results, Implications, and Recommendations 136
Conclusion 162

*Appendixes* 165
*Notes* 193
*Bibliography* 217
*Index* 241

# List of Tables

**Tables**

1 Three Key Concepts  4
2 A Collaborative Framework  49
3 The Presence of Interorganizational Collaboration  84
4 The Quality of Interorganizational Collaboration  108
5 The Barriers to Interorganizational Collaboration  133
6 Summary of Collated Results  137
7 Three Approaches from Three Literatures  166
8 The Relevant Actors  167
9 The Concepts and Their Components  167

# Acknowledgments

Completing a book based on years of research is not unlike disaster response in that both require a high level of supportive collaboration across a dynamic team. The team that took on this challenge with me includes first and foremost my life partner, Mollie Ryan, who endured countless hours of esoteric conversations and delayed travel plans as research monopolized substantial chunks of my time; my parents, Johan Botha and Elizabeth Holland-Muter, both of whom sparked and sustained my pursuit of knowledge as a valuable end in and of itself; Dr. Leslie A. Pal, whose experience and expertise were essential to this project's completion; Dr. Daniel Henstra and Dr. Chris Stoney, who added disaster response to their already busy schedules; and the Social Sciences and Humanities Research Council of Canada, which provided the initial funding support that kick-started my research. The aforementioned, and all the other friends, family, and colleagues who guided and/or endured me throughout this project, formed a team that will always have my deepest appreciation and gratitude.

This book is dedicated to all those individuals across Canada – in and out of uniform – who make it their job to protect their communities during times of disaster.

BOOTS ON THE GROUND

# Introduction

**The Puzzle**

The frequency and impact of many natural disasters has steadily increased throughout the last century and into the current one. Governments, non-profit agencies, the private sector, and international bodies have established organizations with the explicit mandate to help manage natural hazards should they overwhelm a jurisdiction's regular ability to cope. Emergency management (EM) has become "everyone's business."

Federal states like Canada face unique challenges in developing EM systems because at least three levels of government share and/or claim responsibility for managing disasters. In Canada, the challenge of such multilevel governance is (a) particularly acute due to an institutional and historical divide between the federal and municipal levels of government, and (b) relatively unusual given the substantial constitutional responsibility and policy capacity of Canada's provinces. The Canadian constitution "institutionalizes" a federal–municipal divide by entrenching municipalities as mere "creatures" of each province (i.e., they exist exclusively under provincial jurisdiction). This means that while municipal governments are usually the most affected by disaster, EM policy and administration at the federal level – which has the greatest ability to command resources – are not constitutionally linked to municipalities. The provinces, in turn, play a substantial role as this country's constitution sets them up as "co-sovereigns" within a highly decentralized Canadian state. Not only does federal EM support to municipalities flow through the provinces, but the provinces themselves are the key actors during the response phase. A central Canadian agency that oversees the EM of the entire country in the mould of the United States' Federal Emergency Management Agency (FEMA) does not exist.[1]

Table 1. Three Key Concepts

| Research Questions | Concept |
| --- | --- |
| 1. What is the role of CAF in contemporary Canadian disaster response? [descriptive] | The *Presence* of Interorganizational Collaboration |
| 2. How effective is the civilian–military relationship during disaster response? [evaluative] | The *Quality* of Interorganizational Collaboration |
| 3. In what ways might the military contribution to Canadian natural disaster response be improved? [normative] | The *Barriers* to Interorganizational Collaboration |

However, one federal actor that does regularly respond to disasters is the Canadian Armed Forces (CAF). While other federal agencies and departments are engaged during the other phases of EM, CAF is the only salient on-the-ground federal actor in the midst of significant disasters.[2] CAF receives guidance and legal permission from the federal government, is embedded within provincial Emergency Operations Centres (EOCs) as a part of provincial emergency management teams, and performs the actual relief work with municipalities in the disaster zone. CAF is therefore the one organization in Canada that integrates into each level of government during disaster response, yet the effectiveness of such integration has not been systematically assessed.

This book helps to fill gaps in the following areas: the role of CAF in Canadian domestic EM, the nature of civilian–military relations during non-combat natural disaster response, and the study of CAF in public administration. The book addresses a specific puzzle that ties these three areas together: How effectively is the Canadian Armed Forces integrated into the multilevel governance of domestic natural disaster response? This enquiry is broken down into three questions to investigate the descriptive, evaluative, and normative concepts embedded in the enquiry (see Table 1).

As explained in chapter 2, the answers to these research questions are informed by, and have implications for, the interorganizational collaboration and new public governance approaches in the EM and public administration fields, respectively, as well as for the framework of shared responsibility in the civil–military relations literature. This book provides a multilevel governance picture of natural disaster response that focuses on how effectively CAF slots into such governance. The results help identify what works – and what can be improved – in Canada's EM system.

CAF is a regularized yet understudied part of EM in Canada. This book is the first systematic public policy and administration analysis of the nature and quality of CAF's role in domestic EM.

## The Road Map

As this book's "research puzzle" is a function of the nature of Canadian EM and CAF, the first chapter provides an overview of each. Chapter 1 first compares Canada's EM to that of similar federal states: the US and Australia. The historical development of EM in all three countries saw "civil defence" government agencies, originally established to protect communities against anticipated nuclear attacks during the Cold War, expand into all-hazards emergency management organizations (EMOs) by the twenty-first century. Canada is unique in the EM domain, however, given two unusual governance characteristics: weak national disaster response capacity (apart from CAF) with subnational units as the main EM players, and little to no direct federal to municipal support on the EM file. Municipal EMOs therefore vary wildly in capacity across the country while all provincial EMOs enjoy high capacity.

The second half of chapter 1 provides an overview of CAF. The military's underemphasized role in domestic natural disaster response is first demonstrated before its formal mandates, ongoing operations, funding support, organizational features, and bureaucratic structure is discussed. CAF is an intrinsic part of Canadian foreign policy with an array of ongoing operations at home and abroad, despite funding that is consistently below the amount prescribed by the North Atlantic Treaty Organization (NATO). Three branches serve in the air, on land, and at sea, all of which support regional Joint Task Forces (JTFs) that are responsible for disaster responses in discrete geographic areas across the country. Important differences exist between such JTFs, the three branches of CAF, and regular versus reservist units.

Overall, chapter 1 shows that disaster response in Canada is a multi-governmental affair that is complicated by an institutional and historical gap between the federal and municipal governments, and by provinces that have substantial operational capacity compared to their counterparts in other federal states. CAF, an organization developed first and foremost to fight wars, is the one organization that links all three levels of government during disaster response. The nature and quality of this link is demonstrated to have implications not only for the academic literature, but also for emergency management practitioners and policy-makers.[3]

Chapter 2 demonstrates how the research's theoretical framework was developed, lays out the methodology employed, and overviews the selected case studies. Scant academic attention has been paid to the specific research puzzle of how effectively CAF integrates into domestic disaster response. Those disciplines that can guide investigating the research puzzle, given that no specific research on the question exists, are therefore assessed. The EM literature informs the broader assessment of disaster response effectiveness when different levels of government and many organizations respond together; the civil–military relations literature – made up of civilian and military research – informs the understanding of dynamics between armed forces and civilian governments when they work together in a democratic polity; and the public administration literature informs the broader phenomena being analysed when studying governmental disaster response – the implementation of public policy. Chapter 2 uses each of the three literatures' contributions to the general research area of multi-organizational and governmental disaster response to build a framework to address the specific research puzzle of CAF's effectiveness in domestic natural disaster response. While each body of literature differs in focus area, all of them contain contemporary, influential, and significant research niches that identify interorganizational collaboration as the crucial component to achieving desirable outcomes in disaster response, joint military–civilian projects, and public sector initiatives writ large.

Chapter 2 then presents the methodology that can assess CAF's effectiveness during domestic disaster response. Practical and methodological reasons drive the need for a qualitative approach that utilizes case studies of discrete disaster responses. Such case studies have to involve disaster events that occurred after significant changes to the federal EM framework in 2007, represent a variety of hazard types and provincial jurisdictions, and see a significant CAF response. Interviews of CAF officers and emergency management officials, along with archival analysis of each event, provide the data for the case studies. Chapter 2 concludes with overviews of each selected disaster event to (1) demonstrate how each event fits the methodology, (2) share the information the author had when interviewing research participants, and (3) provide the reader with the nature of each disaster event and its attendant hazards, financial impact, and the publicly available information on the actions each level of civilian government and CAF took throughout the response phase.

Chapter 3 is the first results chapter; it provides a descriptive analysis of the presence of interorganizational collaboration before subsequent chapters assess the quality of, and barriers to, such collaboration.

This descriptive analysis addresses the components that constitute the presence of interorganizational collaboration – information sharing, non-manipulative influence, flexibility, support, and collective conflict resolution – in turn.[4] Indicators of all five components appear across the disaster responses. Moreover, the chapter demonstrates that interorganizational collaboration is not only present when CAF plays a role in contemporary Canadian disaster response, but is also a primary feature of such responses. There are other important indicators of the presence of interorganizational collaboration, however, that are not as salient. Hazard type, municipal emergency management capacity, reservist officers, institutional constraints, and specific types of support all play a role in varying the degree of information sharing, non-manipulative influence, flexibility, support, and collective conflict resolution that occur during disaster response.

Chapter 4 is the second results chapter. This chapter moves from the presence of interorganizational collaboration to the quality of such collaboration in order to answer the evaluative question, How effective is the civilian–military relationship during domestic disaster response? This evaluative analysis addresses the components that constitute the quality of interorganizational collaboration – coordination, integration, satisfaction, and trust – in turn. Chapter 4's evaluative assessment of civilian–military effectiveness paints a more complex picture than chapter 3's descriptive assessment of CAF's role. While the majority of indicators of the presence of interorganizational collaboration appear for each component, one component of the quality of such collaboration – integration – sees less than half of its indicators appear. Furthermore, indicators stressed by military research participants as essential to the component of coordination – a common vision of, and agreement on, actions that will lead to a response's end state – are largely absent from the disaster responses. At the same time, all the indicators of the other evaluative components – satisfaction and trust – are prominent features across all events, dovetailing with the descriptive components of non-manipulative influence and collective conflict resolution. The CAF–civilian relationship during domestic disaster response may be characterized as generally effective, but room for significant improvement exists.

Chapter 5 is the final results chapter; it builds on the analysis of effectiveness gaps in chapter 4 by explicitly assessing barriers to interorganizational collaboration, and thereby assessing how the military contribution to Canadian disaster response might be improved. This normative analysis addresses, in turn, the components that constitute barriers to interorganizational collaboration: empire-building, manipulation, distrust, benign incapacity, and conceptual difference. The normative assessment of civilian–military collaboration demonstrates that *intentional* barriers along military–civilian lines are largely absent from domestic disaster response. None of the indicators for empire-building, manipulation, or distrust appear when CAF units respond alongside civilian counterparts. Intentional barriers, however, are not entirely absent from Canadian disaster response – high levels of empire-building are a salient feature of civilian disaster response organizations that exist within the same levels of government.

Chapter 5 shows that the barriers to interorganizational collaboration that do exist along military–civilian lines are due to benign incapacity and conceptual difference. While the indicators for the intentional components of this normative assessment identify contexts where organizational success – implicitly or explicitly – trumps ideal disaster response, the indicators for benign capacity and conceptual difference point to material and psychological realities that constrain response actions even if ideal disaster response is the primary goal for all organizations involved. While benign incapacity affects municipal levels of government and the Royal Canadian Air Force (RCAF) disproportionately, conceptual difference affects all CAF branches and levels of government. With the exception of wildfires, where the nature of the hazard compounds different military and civilian understandings of each other's internal structures and capabilities, all the components of this normative assessment appear evenly across hazards, provinces, and JTFs. As is the case for the quality of interorganizational collaboration, the barriers to interorganizational collaboration are largely shaped by the nature of the organizations themselves, and not by jurisdiction or hazard type.

This book separates the presence and quality of, and the barriers to, interorganizational collaboration in order to analyse the components of each concept in detail as they manifest across different disaster responses. The final chapter then provides an overall picture of how effectively CAF is integrated into the multilevel governance of domestic emergency management during the response phase through (1) assessing all three concepts together, (2) discussing general themes that emerge from such assessment, and (3) recommending how domestic disaster response may be improved.

Chapter 6 thus demonstrates the crucial role of informal networks, the hazard-specific impact of wildfires, the effects of weak federal disaster response capacity, front-line integration as unnecessary collaboration, intragovernmental competition for resources, the formal Request for Assistance (RFA) and "end states" as areas vulnerable to conceptual difference, and the importance of military–civilian "sharing" of decision-making capacity. These themes inform the development of three policy recommendations to improve this country's disaster response system: (1) foster and expand emergency management networks, (2) maintain Canada's decentralized disaster response system, and (3) provide targeted funds to enhance First Nations, municipal, and RCAF emergency management capacity and to support more research on how the all-hazards emergency management model can more effectively incorporate wildfire management during the response phase.

# 1 Emergency Management and the Military in Canada

## Canadian Emergency Management

*One of These Things Is Not (Quite) Like the Other*

In the 1950s, Canada began developing a domestic disaster response system with the goal of managing events that might overwhelm a community's ability to function. At the time, these events were expected to be nuclear attacks sparked by the Cold War, but throughout the last half of the twentieth century a range of hazards began to fall under the system's purview. While not (directly) human-made, floods, wildfires, and ice storms were the type of "non-routine" emergencies that fit the assumptions of the 1950s system. Like nuclear attacks, these natural hazards were generally anticipated, but their timing, location, and potentially disastrous impact on lives and property could not be exactly determined. Canada's governmental EM system, and the agencies tasked with operationalizing the system, therefore gained an "all-hazards" focus during the late twentieth century (Scanlon 1995). By the twenty-first century, all-hazards EM had become a professionalized field with a presence in the bureaucracies of federal, provincial, and municipal governments.[1]

The transition from nuclear attack–focused EM to an all-hazards approach is mirrored in comparable federal states, namely the United States and Australia. The US's mid-twentieth-century EM development largely reflected Canada's. After the Second World War, nuclear attack was perceived as the most serious large-scale domestic threat in the US, which – like in parts of Canada – engendered the term "civil defence" into the titles of agencies mandated to mitigate such attacks. As the intensity of the Cold War declined and wildfires, floods, tornados, and hurricanes replaced war as the main perceived "non-routine" threat to American lives, natural hazards began to fall under the purview of "civil defence" agencies (Knowles 2011). Australia saw the same development in the perception of hazards, from one initially focused on national security to one including natural threats as the country moved further away from the Second World War. Voluntary civil defence associations – organizations largely disbanded in Australia after the Second World War – were resurrected in the wake of the Cold War, and ultimately became focused on droughts, cyclones, wildfires, and floods as each Australian state began dealing with these hazards (Jones 2007).[2]

Despite Canadian, American, and Australian similarities in hazard perception throughout the twentieth century, Canada's history and institutional structure distinguishes its public administration of EM. Canada's "subnational" units – the provinces – do the bulk of EM while the federal government's focus on EM ebbs and flows (McEntire and Lindsay 2012; Juillet and Koji 2013). While Canada's constitution does not explicitly allocate EM to a particular level of government, the decentralizing pressures in the country's legislative framework (Malcolmson and Myers 2012), along with Quebec's historical insistence on greater independence (either within the federation or as a sovereign state), has meant that managing hazards, which are most often regional in impact to begin with, has become a largely provincial responsibility (Lindsay 2014). The United States and Australia have avoided placing a similarly disproportionate level of EM operational capacity on their states (the equivalent to Canadian provinces). While the US provides greater disaster response decision-making power to its federal level, both the US and Australia see EM capacity distributed more evenly across their state- and municipal-level governments compared to Canada.

By the time Canada began redirecting its historical tendency of focusing intergovernmental affairs on federal–provincial relations to the detriment of federal–municipal relations (Stoney and Graham 2009), the growing importance of cities as prime actors in the public policy process was already established in the American context (Miller and Cox 2014). US attempts to improve its disaster response systems included municipalities at the decision-making table by the early twenty-first century. Strong critiques of the now replaced American Federal Response Plan identified a lack of sufficient influence from local governments (Kapucu 2009a; Harrald 2012), and subsequent federal policies brought municipalities in to address these critiques. Cities provided locally influenced solutions to EM problems and the collaborative results were positive (Hu, Knox, and Kapucu 2014).

Australia, in turn, has a legally loose federal EM framework to allow the lower levels of government to tailor their disaster response. While this has raised questions around the lack of standardization of disaster declarations, Australia's approach affirms the role of local governments, businesses, and non-profits in not just the immediate response phase, but throughout the emergency management cycle (Attorney General's Department 2009). Australia's encouragement of local government EM capacity is buttressed by a deep volunteer sector that – unlike in the US and Canada – is largely proactive and mitigation-focused in its disaster-related work. Indeed, Australian municipalities have an institutionalized citizen-based EM community to draw upon not just as financial donators, but as human resources in preparation for, response to, and recovery from disasters (Peters and McEntire 2010).

Meanwhile, direct federal–municipal relations on the EM file in Canada are minimal. The institutional and historical reasons hinted at above can be elaborated to explain this dynamic. As the Canadian constitution "institutionalizes" a federal–municipal divide by entrenching municipalities as mere "creatures" of each province, federal–municipal policy dialogue on EM, whether on mitigation efforts, response plans, or recovery funds, is "filtered" through the provincial level. Political parties and provinces have used this institutionalized divide to protect their interests when debating EM legislation. During the 2007 federal debate on the *Emergency Management Act*, sovereigntist Quebec MPs refused to allow provisions for direct federal–municipal relations in the Act while federalist government MPs found it politically easier to acquiesce to demands for provincial jurisdictional protection, regardless of what EM subject matter experts had to say on the subject (Juillet and Koji 2013). While the Canadian government spent the first decade of the twenty-first century revamping *federal* EM policy (the provinces were already established as EM leaders), little was done to address the divide between federal and municipal governments. The importance of federal–municipal collaboration is affirmed throughout EM research,[3] but historical and institutional factors appear to direct the Canadian public policy process away from such best practice.

The disproportionately provincial focus on EM has been somewhat tempered in recent years by two related developments. First, the 9/11 terrorist attacks against the United States saw EM conceptualized as a policy area relevant to national security, which is firmly ensconced in federal jurisdiction. This link to the Government of Canada and a suddenly super salient policy sphere helped provide a gateway towards the second development: the revamping of more general federal EM policy-making throughout the first decade of the twenty-first century. These policies, as most prominently manifested in the 2007 *Emergency Management Act*, showed concerns with a range of hazards sparked by climate change as opposed to only security threats (Juillet and Koji 2013). The impact of wildfires and floods drew special attention as they are Canada's most frequent natural hazards in the summer. However, the same institutional and historical forces mentioned above curtailed the transformational nature of federal EM policies. The growing understanding of EM as a policy domain that requires collaboration among all levels of government was not powerful enough to diminish the incentives for instrumental actors – mainly, but not only, Quebec MPs – to leverage Canada's constitution, which defines municipalities as purely a function of provincial legislation, in pursuit of political, not EM, goals.

The nature of Canadian EM is therefore unusual for two related reasons: First, a key characteristic that defines Canadian EM – the extensive role of subnational units – does not exist to the same degree in comparable states, and second, a characteristic crucial to effective EM in those comparable states – robust and direct federal–municipal collaboration – is minimal in Canada. The one part of Canada's EM system that maintains a consistent link between the federal and all other levels of government during disaster response is CAF when it responds to domestic disasters.

## "Coordinating" Disaster in Canada: Managing the Unmanageable

Organizations tasked with EM come under a variety of slightly different titles. The term "civil defence" had largely disappeared in Canada, Australia, and the United States by the end of the twentieth century as some variation of "disaster/emergency management agency" became widespread. Most emergency management government organizations in Canada currently fall under the title "Emergency Management (or Measures) Organization (or Agency)" (Government of Canada 2015). The recent rise of similar organizations housed in policy-specific departments, such as "health emergency management" branches, or of organizations that include a specific hazard in their title, such as "fire and emergency services," has somewhat complicated such generality. However, while some organizations may be biased towards a specific hazard, which can occur when membership is largely made up of individuals who in their former, pre-EM careers mainly trained to mitigate a specific hazard or set of hazards (i.e., firefighters), the organizational EM model developed in North America follows an all-hazards approach on paper (Rodriguez, Quarantelli, and Dynes 2007; Waugh and Tierney 2007). Unlike the model where the lead coordination agency varies depending on the hazard (Haddow, Bullock, and Coppola 2013), a single organization – or in the case of small municipalities, an individual – exists at each level of government in Canada that carries a general coordination mandate. Policy-specific organizations like health emergency management branches manage disasters for their policy domain only, and are not responsible for a whole-of-government coordination effort, but still apply the all-hazards model within their policy domains. For the sake of convenience, all organizations mandated by government with the coordination of other organizations during disaster response will be referred to as emergency management organizations (EMOs), regardless of the level of government or scale at which they function. Policy domain–specific coordination agencies will have their policy area mentioned in their title (and are included in this book because they are civilian EM-related agencies that engage with CAF during disaster response).

EMOs should not be conflated with Emergency Operations Centres (EOCs). EMOs are organizations while EOCs are physical sites from which a specific event is managed. EMOs have an evolving, but constantly active, organizational life, while EOCs activate for the duration of a hazardous – or potentially hazardous – event, or serve as a glorified meeting room for emergency management stakeholders. An EMO will have an EOC, but an EOC does not require an EMO. Many disaster and emergency response organizations may have an EOC, including front-line organizations like police departments, but may not have the EMO-distinguishing feature of a specific mandate to coordinate other organizations during disaster response. EMOs should also be distinguished from "business continuity" services found in private and public organizations. The overarching goal of business continuity planning is the survival of their organizations as an end in itself, while the overarching goal of an EMO is to coordinate those organizations that respond to a disaster that overwhelms an entire community. Business continuity aims to keep a specific organization resilient, while an EMO aims to keep people and property within a defined jurisdiction resilient (Federal Emergency Management Agency 2018). Some organizations, academic literature, and professional standards conflate the EMO and business continuity function as many of their professional practices overlap,[4] but this book distinguishes between the two functions and focuses on the former.

The different levels of Canadian government need to be understood to appreciate the unique way EM manifests within each. The municipal level of government is one of the more under-studied and under-theorized areas of governance in Canadian public policy, public administration, and political science, and as such is the most difficult level of government to define. A growing appreciation for the role of

Canadian municipalities in the public policy process gained traction in the last decade of the twentieth century (Andrew 1994, 1995), which joined a concerted American focus on urban governance by the twenty-first century (Miller and Cox 2014), including assessments on how municipalities fit more robustly into multilevel governance (Stoney and Graham 2009). However, by definition a municipal government continues to be any legal jurisdiction that (1) gains its legitimacy through provincial legislation and (2) creates and enforces laws and policies for a sub-provincial jurisdiction that is constrained by provincial and – indirectly – federal laws and policies. The former corresponds to what constitutionally defines municipalities in Canada (i.e., "creatures of the province") and the latter's law-making criteria distinguishes municipalities from organizations that also gain legitimacy through legislation, whether airport authorities or Crown corporations. The two criteria that define municipalities do not say anything about jurisdictional types or resource capacity. This means that overlapping sub-provincial jurisdictions (i.e., a city that exists within a county) and small/weak sub-provincial jurisdictions (i.e., a small town that is largely a "policy-taker" from a regional government) all count as municipal governments. Such a wide net is relevant to understanding municipal EMOs because disasters do not discriminate according to local government type; whichever local government is closest to a disaster's initial impact will likely be affected first regardless of its response capacity.

The capacity of municipal EMOs differ significantly depending on a municipality's size, regional economy, provincial–municipal relations, history with disaster, proximity to natural hazards, and the salience of EM in the municipality's policy process. For example, the City of Calgary's EMO is so robust in terms of staff and facilities, which includes a state-of-the-art EOC, that it handled the disastrous 2013 multi-river flood largely by itself. Despite less regular experience with floods than its fellow prairie province Manitoba, Alberta's comparative wealth, and an EM structure that is a function of the province's broader public administration structure, which devolves a comparatively large amount of responsibility and capacity to cities, means powerful large city EMOs (Hale 2013). By contrast, the City of Winnipeg – despite significant experience with floods – sees a substantial role played by the provincial hydrologic forecast centre and EMO in managing floods that threaten the city. This is partly due to the important historical role the province has played in keeping the city that holds more than half of the province's population safe from floods (Haque 2000). Indeed, the City of Brandon, despite much humbler general resource capacity compared to the provincial capital of Winnipeg, largely managed the 2011

Assiniboine River flood without significant provincial support. Canada's largest cities of Toronto, Montreal, and Vancouver all have EMOs and EOCs that correspond to their size, while Ottawa's dedicated EM staff has often been less than ten individuals (Willing 2017), perhaps because the country's capital, unlike other Canadian cities, is saturated with EM-like government organizations focused on keeping the National Capital Region safe and secure. Cities east of Quebec tend to rely heavily on provincial EM support and have correspondingly smaller EMOs (with Halifax as a notable exception).

The aforementioned institutional factors of Canadian federalism also play a role in determining the scope and capacity of a Canadian city's EMO. Municipalities that are not a provincial capital or one of the largest cities in the country can struggle to enhance EM clout, even if they do have the makings of a robust EMO. During the 1998 Ontario ice storm, the then existing Regional Municipality of Ottawa-Carleton (RMOC) clearly had the most resources compared to other affected municipalities. Its EOC was functioning shortly after it was clear that the storm would become a disaster, and communication channels with CAF were established early (Scanlon 1998). The RMOC could not, however, horizontally capitalize on its capacity to help other municipalities because of the amount of power the Canadian constitution grants to provinces. The RMOC was willing and able to help, but other regions perceived Toronto as the legitimate option for aid while the provincial EMO inhibited the RMOC from helping in areas beyond its own region. The province even constrained the military's ability to act. Since health care and health care providers are provincially run and regulated, Ontario stopped CAF medics from providing medical services, despite their high level of training (Scanlon 1998). These – arguably negative – outcomes for disaster response had little to do with municipal resources, expertise, communication, or leadership capacity. Rather, the inhibition of the RMOC's EMO was a consequence of a constitutional structure that prioritizes the power of the provinces.

Despite the strong EMOs in Canada's largest cities and the occasional robust – albeit constrained – EMOs in mid-sized municipalities, there is significant variation in EMO capacity, particularly among smaller municipalities. Most small towns, counties, and rural municipalities have "EMOs" of one person, and EM often exists as but one responsibility among that person's many other duties. Furthermore, while the formally mandated EMOs in large cities are the organizations that do, in fact, pick up the mantle of coordination during a disaster, a discrepancy exists in small municipalities between the individual formally mandated with EM and the individual(s) who takes the lead when disasters strike. It is common for a local fire chief, popular mayor, or other important community member to take on the coordination function regardless of who formally is responsible for and familiar with the function. The variation in municipal EMO capacity means that CAF encounters vastly different response capacities, in terms of levels of effective municipal resource coordination and the amount of resources themselves, when working with municipalities during disaster response. CAF cannot plan for a standardized capability when arriving at the front lines of a disaster with a municipal partner.

Provincial governments, in turn, are the geographically discrete political jurisdictions that are imbued by the Canadian constitution with responsibility over a range of important policy areas. Canada is one of the world's most decentralized federal states, rendering its provinces more "co-sovereigns" than mere subnational units (Malcolmson and Myers 2012). The implications here for EM is that quasi-state jurisdictions with significant taxation power, exclusive control over many policy areas, and a constitutional monopoly over every municipality within its borders face both a demand from their citizens for EM services as well as an incentive to pursue EM activities to benefit the power of the province.[5] Some EM services are natural extensions of provincial responsibility, such as maintaining the integrity of essential infrastructure, protecting sectors crucial to a provincial economy, and ensuring that health care services continue to function. The federal government may directly or indirectly support any of these initiatives through funding, but the implementation is up to the provinces. Hence the operational primacy of the provinces within the EM domain.

Politically, the "co-sovereign" status of provinces places significant responsibility for EM success or failure on provincial leaders. Rural leaders in Canada can easily point to lack of support – financial or operational – from higher levels of government. Within a province's sphere of responsibility, however, there is no higher level of government. While the federal government can be blamed for not providing enough disaster recovery funds or not adequately supporting mitigation efforts, failures during the response phase land squarely on provincial shoulders. Even CAF is sent in only by provincial request; the federal government does not unilaterally send in troops to aid the public. Indeed, there are relatively recent historical examples where provinces appear to have at least in part initiated CAF resources themselves, as the Province of Ontario did with the first CAF resources out of the gate during the 1998 ice storm (Scanlon 1998).[6] Furthermore, CAF can be ready to deploy anywhere domestically within mere hours; the formal Request for Assistance (RFA) process can be streamlined so as not to delay troops, and the federal government has never refused a request for troops during a significant disaster. In other words, if there is a delay in deployment of that distinctly federal resource (the military), the province is still to blame. Indeed, during the 2017 western Quebec floods the province kept CAF deployed even as water levels receded, in part to respond to criticism that CAF had not been requested by the province soon enough (Smith and Marandola 2017).

While the inherent complexity of a disaster event allows for accountability shirking or "blame avoidance" if more than one actor is responsible for outcomes (Moynihan 2012), Canadian provinces are often the one actor responsible during disasters with nowhere else to point should the response go wrong. Provincial premiers are therefore eager to appear in control during a disaster for as long as possible, which – ironically – is one of the reasons requests for federal assistance in the form of CAF can be delayed until significant damage is already done, harming the premier's reputation in the process.[7]

The bureaucratic result of the role the provinces play in Canadian disaster response is provincial EMOs that are large and sophisticated compared to their municipal and federal counterparts. When province size is considered, provincial EMOs do not vary in capacity to the degree that municipal EMOs do. Each provincial EMO is enacted by EM-specific provincial legislation, ensuring a level of salience, proximity to cabinet ministers, and funding that other branches within regular line departments rarely enjoy.[8] Furthermore, the doctrine of ministerial responsibility ensures that the minister responsible for EM takes criticism after EM failures, while the EMO itself is perceived as an organization that requires more support after disaster response failures. Indeed, while disasters can be detrimental to political careers, they are good business for EMOs: "[More] disasters mean the need for more [EMO] budgets, more manpower, and eventually more recognition" (Kirschenbaum 2004, 99).

The increased frequency and impact of natural disasters due to climate change has prioritized provincial EMOs in relation to many other bureaucratic branches and agencies. Provincial EMOs, despite being units within government departments rather than existing as departments in and of themselves, have a cabinet minister assigned to them. They have their own EOCs, have central staff in a province's capital as well as staff in provincial regions, and tend to attract hires with significant experience such as retired senior military officers and firefighters.

Variation does exist for provincial EMOs in how they fit into each provincial bureaucracy. They might be nestled in with firefighting agencies (Newfoundland and Labrador), share a minister with municipal affairs

(Alberta), sit in an infrastructure department (Manitoba), or fall under a province's more conventional public safety portfolio (British Columbia). Regardless of a provincial EMO's bureaucratic location, CAF can expect to work with a relatively well-resourced and trained organization when responding to a domestic disaster alongside any province in the country.[9]

Finally, the federal government is the single level of government responsible for constitutionally defined areas of national concern. The EM function at the federal level is therefore linked to country-wide public safety initiatives, but is also largely a corollary of national defence. The responsibility to keep the country safe from harm goes beyond sending troops overseas to pursue Canadian interests; the federal government becomes constitutionally implicated in a domestic disaster should the integrity of the Canadian state be threatened. While this could mean deploying CAF to respond to human-caused threats, it has become common to deploy CAF when a crucial part of the Canadian state – a province or territory – is overwhelmed by a natural disaster.[10]

The civilian federal EM apparatus is led by the federal department of public safety. The mandate of Public Safety Canada is to protect Canadians from a range of risks that include natural disasters, terrorism, and crime (Public Safety Canada n.d.). The department was established in 2003 to facilitate coordination across those federal organizations tasked with ensuring national security and the safety of Canadians. This broad mandate informs the department's more specific raison d'être and responsibilities (Goodale 2016). The following are the three key roles of Public Safety Canada as established in legislation:

1 Support the minister in public safety and emergency management programs not overseen by other federal departments.
2 Demonstrate national leadership on the national security and emergency preparedness files.
3 Support the minister in coordinating those entities that fill the public safety portfolio.

Note that the first of these functions of Public Safety Canada allows for other federal departments to shoulder a significant degree of EM responsibility (a potential distribution of EM roles not seen beyond EMOs in provincial bureaucracies); the second stipulates "preparedness" as opposed to "emergency *management*," which would indicate the entire EM cycle of – at a minimum – mitigation, preparedness, response, and recovery; and the third is not an EM function at all but rather is focused on the internal management of the department.

Public Safety Canada's responsibilities are numerous, with the Emergency Management and Programs branch competing for resources and attention against other salient branches such as National and Cyber Security, Portfolio Affairs and Communications, and Community Safety and Countering Crime. Furthermore, organizationally the Public Safety Portfolio is composed of one "umbrella" department that oversees an array of entities, none of which focus on EM.[11] Unlike those entities established as "agencies," such as the Canada Border Services Agency (CBSA), the Canadian Security Intelligence Service (CSIS), and the Royal Canadian Mounted Police (RCMP), the EM function of Public Safety Canada is not contained within its own agency; it sits within the umbrella department. In other words, the same entity responsible for managing all the non-EM public safety–related entities is also charged with being the federal EMO.

The Canadian federal EMO, then, is a minor organization compared to the Federal Emergency Management Agency (FEMA) in the US, which is active in all phases of EM, is explicitly mandated with EM and not with intradepartmental management, and is tasked with the coordination of all organizations and resources should local and state capacity for response be diminished. After the 9/11 terrorist attacks, FEMA was placed under the auspices of the Department of Homeland Security but remained its own distinct agency of substantial scope and size, with its own identity and evolution within the US federal government (Waugh and Tierney 2007; Cigler 2009). Furthermore, while FEMA plays an influential role in the standards, guidelines, and funding of regional emergency management,[12] Public Safety Canada has not been a salient presence in EM, which is largely handled by the provinces through their own legislation (Lindsay 2014). In the twenty-first century this may be due in part to the fact that no national disaster has ever been declared under the 2007 *Emergency Management Act* (Lindsay 2014).

The implication of the above is that the main link between Public Safety Canada and CAF is largely constrained (1) to the RFA process, when the ministers of Public Safety Canada and the Department of National Defence jointly allow the domestic deployment of troops; and (2) to EOCs, where CAF liaison officers and Public Safety Canada representatives are both present. While on paper Public Safety Canada "clears" the use of troops as a federal resource and acts as a link to the federal government writ large, the operational EMOs that CAF works with during disaster response are mainly provincial and municipal. Indeed, a senior federal bureaucrat with extensive experience on the federal EM file, including experience liaising with CAF during disaster responses, characterized the federal Government Operations Centre (GOC) as merely an "information clearing centre," and the EMO function of Public Safety as a "toothless paper tiger," meaning it has no real operational capacity.[13]

The federal EM capacity as elaborated above holds true in almost every circumstance except during relatively small disasters that are geographically constrained enough to only significantly impact a First Nations community or communities. As in other policy domains, the federal government is responsible for disaster response in First Nations communities in a way that has no corollary to other communities such as municipalities, which are the "legal creatures" of each province. Disaster response by the federal government in First Nations communities is led not by Public Safety Canada, but by the federal department(s) specifically responsible for service delivery in First Nations communities (most recently Indigenous Services Canada). The model for such disaster response takes a variety of forms, but very rarely does the federal government itself take on a type of operational role that reflects provinces supporting municipalities. Indeed, apart from a few exceptions – such as Indigenous Services Canada contracting out the Red Cross for First Nations in Manitoba – the federal government often contracts out provinces themselves to support First Nations communities.

Furthermore, while the complexity between First Nations communities and the federal government on the EM file warrants its own analysis, the very nature of large-scale disaster response diminishes the impact of such complexity as (1) the resource constraints of individual First Nations communities sees them largely respond similarly to municipalities, and (2) the resource capacity and operational clout of the provinces, along with the functional logic of disasters that prioritizes such capacity and clout, means that provinces remain the key suppliers of disaster response actions to essentially all communities within or overlapping their borders, including First Nations. CAF officers certainly have to be attuned to the fundamental distinctions between First Nations communities and municipalities, with nationhood and sovereignty relevant to the former but not the latter, but during large military deployments to diminish a natural disaster they will see the same varying EM capacity in First Nations communities as in municipalities.

While the nature of CAF will be discussed at length in the following section, it is important at this juncture to note why CAF cannot be considered a mere function of federal government paradigms and policies (i.e., why it fills a distinct space within the multilevel governance of disaster response). The two main reasons are the institutional separation between CAF and the federal government, and the operational reality of disaster response. While CAF is an instrument of the federal government and the ultimate symbol of the federal government's alignment with the Canadian state writ large through its capacity to monopolize the use of violence, CAF is not like any other federal department or agency. Unlike line departments, including its "sister organization," the Department of National Defence, the administration of CAF is internally controlled with minimal influence from trend- or standard-setting bodies in other parts of the federal government. Organizational structure, movement through that structure (i.e., channels of promotion), and generally the unique institutional culture – which shapes the incentives of individuals within the military – are all jealously guarded (Granatstein 2002). Indeed, unlike other government employees (and the civilian Canadian population at large), CAF employees are governed by unique extensions to legislation that are enforced by a separate military judicial system. Furthermore, while a range of relatively independent "special operating agencies" exist in the federal government, from the RCMP to the Communications Security Establishment, none of them are given the discretion that military officers receive when managing internal affairs. The link between an entry-level policy analyst to their deputy minister – tenuous as it may be – is far more direct than the link between the new soldier recruit and *their* federal deputy minister at the Department of National Defence.

Historically in Canada the military has responded to "fill a void" in disaster response without first receiving confirmation and instruction from the federal government. During one of Canada's – and, indeed, the world's – most destructive industrial accidents, the 1917 Halifax harbour explosion, soldiers on-site spontaneously formed up and initiated a response (Scanlon 2005). More recent disasters, such as the 1998 Ontario ice storm, saw provincial calls for help directly to the relevant regional military authority, which took its first actions without hard federal government confirmation (Scanlon 1998).[14] While formal procedures for military involvement are established, the operational reality of disaster response – especially in sudden, dire events with lives and property on the line – will not see a regional brigade or local regiment wait for official confirmation from Ottawa before responding to local calls for help. Indeed, a legal pathway exists for commanding officers to order their troops to save lives and property without the provincial RFA in place.[15]

A full overview of Canadian EM requires a final note on civilian EMOs and disaster response coordination. Recent analysis of North American disaster case studies since disaster research formally began in the 1950s has shown that desirable disaster response outcomes can occur in the absence of formal coordination, and in some cases tentative links can be made between the nature of coordination agencies and undesirable disaster responses (Botha 2018). The impact of EMOs on disaster response outcomes is difficult to measure given ongoing confusion around what, exactly, EM as manifested in a coordination agency entails. Despite the formal mandate of EMOs and the types of events they generally address, Schroeder, Walmsley, and Ward (2001) note the following:

> [We have not] completely settled how emergency management should be organized ... [There] are seemingly intractable problems of organization, administration, and coordination. How can one agency be given the power and jurisdiction necessary for effective disaster planning and coordination of response and recovery operations without giving it more power in times of both non-emergencies and emergencies than other participants in the political process are willing to grant it? (359)

The authors here tap into the political problem of power sharing, policy agendas, and empire-building inherent to the public policy process. This problem suggests that high value issues will be drawn to the most powerful actors in the process, which is indeed what happens when a disaster reaches a certain scope. The process for declaring a disaster, and thereby denoting what "counts" as a significant adverse event, is a political process, not one based on consistent criteria (Cutter 2005). The location of EMOs in bureaucracies is affected by political preference (Cigler 2009), and leadership performances during disasters demonstrate the direct link between high-ranking politicians and the emergency management file (Kapucu 2009b; Solomon 2017). Furthermore, the ability to perform effective interorganizational coordination can be a function of how tightly an emergency manager is linked to "the key point of authority and power" (Drabek 2010, 217). In his classic multi-year comparative study on the effects of centralization in the United States, Japan, and Italy on disaster responses in the mid-twentieth century, McLuckie (1977, 78) noted that the final authority for coordinating a response during a major disaster automatically moves to the relevant political authority.[16] On a smaller scale, Fritz, Rayner, and Guskin's (1958) field-establishing case study on behaviour in an emergency shelter during a North American snowstorm demonstrated that the coordination function automatically moves to individuals who arise as political authorities. The problem these dynamics pose for assessing the impact of EMOs is that the moment adverse events reach a point where they need "professional" coordination they are often salient enough issues for the political authority to take over, hindering the coordinating authority an EMO might have had.

Schroeder, Walmsley, and Ward (2001) also point to an operational problem: Even if granting EMOs coordinating power during the response phase of a disaster was universally accepted, is it possible for one agency to effectively coordinate the multitude of organizations involved? Considering that these organizations might include the Red Cross, the Salvation Army, other agencies in government, police and other emergency services, and the military, effective coordination is far from obvious. Indeed, lack of coordination is identified as a perennial problem for public administration across the board (McGuire 2006), and a key insight from the study of complex systems – as disaster response systems certainly are – is that no single agency possesses the capacity to manage large-scale threats (Skertich, Johnson, and Comfort 2012). As just another creature of bureaucracy, it is not clear how an EMO – despite its official mandate – should solve a problem that transcends its purview and capacity.

The limits of formal coordination are important to keep in mind when assessing CAF's integration into domestic disaster response. The investigation into how effectively CAF contributes to the Canadian disaster response system requires an appreciation of the fact that CAF's main partners in disaster response are agencies formally mandated with coordination, a task that by all accounts faces an array of intractable problems. CAF is tasked to work with agencies that have been given the responsibility of coordinating the entire response; how this dynamic plays out will be illuminated in the results chapters.

## The Definitions of "Disaster" and "Emergency Management"

The final piece in providing an overview of the Canadian EM system requires a discussion of how disasters and emergency management are defined by those who work in and study EM. The definition of "disaster" has been intensely debated in the academic literature. One of the few dominant threads characterizes disasters as adverse phenomena above and beyond the "normal" or regular processes that a given social system usually accommodates. The mid-twentieth century – the "classical period" of North American disaster research, dominated by sociologists – saw disaster definitions that stressed an "agent as catalyst (hence the use of the term "event"), [but] most really dealt with the social disruption attendant to the cause rather than the cause or agent itself" (Perry 2007, 5). In other words, disasters disrupt social order in a manner that requires people to cope by leaving patterns of "norm expectations" (Killian 1956, 67). Key features of disasters include the need for *new* behaviour patterns to be adopted (Moore 1958, 310), mainly because during disasters "essential functions of the society [are] prevented" (Fritz, Rayner, and Guskin 1958, 655). The cause that triggers such prevention was characterized as external to the social system *and* not subject to control (Sjoberg 1962, 357). Despite shades of difference between the early disaster scholars, all acknowledged the conceptual necessity of elevating disasters to phenomena above and beyond those adverse events regularly accommodated by a given social system. Resource limits, communication breakdowns, or poor individual leadership are relevant to disasters only when they contribute to a system-wide vulnerability. They are not disasters in and of themselves.

As public administration scholars, emergency management professionals, engineers, meteorologists, geologists, and others joined the discourse on what a disaster entails, the number of definitions exploded as individuals employed definitions that were most useful to their own academic or practical concerns (Perry 2007). Since this book focuses on the implementation of public policy by government actors, the definition of disaster will be operational and legal, not sociological, in nature. Disasters are here understood as those "non-routine" emergencies that fall under the purview of EM legislation and policies and are managed by organizations with an explicit EM mandate. Note, however, that as in the classical definitions, the theme of "normalcy disturbed" is maintained: Non-routine emergencies are events that, while "generally anticipated, and for which there may be generic plans, [always] stretch the emergency system, and require some shifts in operational procedures and thinking through more than expected scale, complexity and/or uncertainty" (Handmer and Dovers 2013, 107). Examples include large fires, major storms, intense flooding, epidemics of known diseases, and multi-vehicle accidents. In legislative terms, "non-routine emergencies" dovetail with the events anticipated in the 2007 federal *Emergency Management Act* and each province's corresponding legislation.[17] Violent events deliberately caused by humans, such as terrorist attacks, are purposefully not listed above. The public administration and academic study of security is beyond the scope of this book for three main reasons: (1) Human behaviour differs significantly during violent disasters compared to natural ones (Drabek 2010); (2) CAF requires extra and separate legal permission to assist law enforcement in any way (Canadian Joint Operations Command 2015); and (3) part of what makes this book's enquiry interesting is its focus on an organization – moulded primarily to apply legal force – that is tasked to function in a non-combat context in concert with non-security related civilian agencies.

There are two other broad categories of "disaster events" that do not fall directly under the purview of EM and are excluded from this research.[18] "Routine events" are lower in intensity and higher in frequency than non-routine events, and are largely handled by hazard-specific and front-line emergency response organizations such as the police, fire, and emergency medical services. "Complex events," in turn, such as the impacts of climate change or severe and widespread socioeconomic decline, transcend any agency and demand the consistent attention of the highest political authority. Responses to complex events generally entail the entire social, cultural, economic, and political system.[19] While EM legislation, policies, and actors can be involved in routine and complex events, they are the main players in non-routine events.

The definition of "emergency management" itself has seen less debate. EM scholars, practitioners, and legislators in Canada and the US have defined EM as the preparation for, prevention and mitigation of, response to, and recovery from emergencies and disasters. This definition can be made more specific by characterizing "emergencies and disasters" as adverse events that *transcend routine emergencies*. This addition allows the EM definition to dovetail with the classical definitions of disaster by focusing on vulnerable points of a given system rather than including almost all adverse events, many of which may be accommodated by the regularized operations of a system. Indeed, Canada's *Emergency Management Act* implicitly assumes that "emergency," in the context of the Act, is understood as an adverse event that transcends "regular" and expected emergencies. Emergency medical services personnel and first responders (i.e., police, firefighters) are the regularized operators within the system that respond to and hence ameliorate such emergencies. These actors gain their authority from actor-specific provincial and/or municipal laws, and not from EM legislation. The very existence of federal and provincial EM legislation above and beyond laws that establish general emergency services implies an additional conception of adverse events and the organizations meant to lead responses to them; that is, events and organizations that are different from those that spark and respond to the average 911 call.

CAF only slots into the disaster response system when a non-routine event has transcended a province's ability to manage the event, and where the main actors are the provincial and municipal EMOs (and their on-the-ground representatives and corresponding political masters). Police, fire, and medical services are, of course, involved in such events, but they do not make the strategic calls and are not CAF's main partners in response.

## Calling in the Cavalry: A Primer on CAF and Military Responses to Disasters

### A Historical and Comparative Perspective

Despite the perception that militaries in liberal democratic countries are called upon for domestic operations only in extraordinary circumstances, the armed forces of Canada and comparable countries have long been involved in domestic disaster response. In the 1917 Halifax harbour explosion – one of Canada's earliest and worst disasters – members of the Army and Navy played a main role in coordinating and enacting the city's response (Scanlon 2005). Indeed, essentially all national governments use their militaries to help manage disasters (Shemella 2006). This aid brings militaries close to those levels of provincial and municipal governments that have no legal power over armed forces and little regularized experience in working with them (Skertich, Johnson, and Comfort 2012; Juillet and Koji 2013).[20]

The rise in the frequency and impact of natural disasters due to climate change, along with terrorism as the most salient twenty-first-century security threat, renders outdated the conventional understanding of the military as a tool used mainly in state-on-state warfare. The roles and missions of armed forces have been more dynamic since the end of the twentieth century than ever before (Arquilla, Ronfeldt, and Zanini 1999). Indeed, Shemella (2006) notes the following: "The key question for governments is not whether to use military forces at home but whether to assign each duty as a role or mission. If military support to civilian authorities become systematic, it should make the transition from being a mission to being a role" (135).

While no academic study has attempted to define CAF's impact on domestic EM as a "role" or a "mission," an overview of the Department of National Defence (DND) mandate and its annual performance reports to Parliament strongly suggest that CAF does play a "role" in EM. Not only does DND consistently prioritize domestic humanitarian "emergency preparedness" in its core mandate, but CAF responds annually to domestic disasters that have overwhelmed a community's ability to function (DND 1997–).[21] These responses are to hazards that range from wildfires in Western Canada, ice storms in Central Canada, hurricanes on the East Coast, and floods across the country. While the media reports on the most significant of these hazards, the amount of media attention – usually coverage of a particularly salient event that fades fast after the main response (Birkland 1997; Scanlon 2007) – does not correspond to the much broader extent to which CAF responds to disasters.

## The Mandate, Organizational Features, and Bureaucratic Structure of CAF

While the extent of CAF response to natural disasters does suggest that such responses have become one of CAF's key roles, most CAF resources are still spent on national defence activities against perceptions of human threat. The range of tools and skills CAF uses during domestic disaster response are therefore largely convenient by-products of tools and skills required to fight wars. Even the most civilian functions provided by CAF during disaster response, such as Army engineers building bridges, is a function that is learned first and foremost as a combat function (to provide efficient transport of troops and materiel).[22] Infanteers clearing debris happen to have significant upper body strength from extensive training holding rifles and machine guns aloft for long periods of time to focus on, or consistently fire upon, a simulated – or, in the case of deployments, actual – human targets. Hercules aircraft can efficiently carry an array of resources because they were designed to carry almost anything contemporary battles may require. Perhaps most to the point, the efficient organizational structure that allows the military to quickly accomplish difficult tasks of substantial scope is a structure that exists because it has been developed for that especially complex and dangerous human activity – the systematic application of violence to destroy a defined enemy. Every single soldier who deploys to fight a natural hazard in Canada has gone through Basic Military Qualification, which means they have been trained on how to use CAF's basic rifle, the C7. Furthermore, most soldiers who deploy during natural disasters hail from the combat arms, which means that the most visible uniformed individuals lifting sandbags, fighting fires, clearing debris, checking in on evacuees, measuring water levels, handing out water bottles, and so on, are trained in a range of weapon systems, as well as hand-to-hand combat.

While troops receive extensive information before deploying domestically, including the tasks to which they are limited (from their superiors),[23] information on the community they are supporting (from liaison officers), and media awareness training (from public affairs officers), no military trade exists that is specifically developed to produce non-combat, domestic emergency responders/managers. In other words, CAF is not a glorified civilian emergency services organization;

it is an organization developed primarily for fighting wars made up of members whose skills happen to be very useful during a natural disaster. The assessment of how effectively CAF contributes to domestic EM therefore requires a broader understanding of the organization and its contemporary context.

CAF is made up of approximately 68,000 regular force and 28,500 reserve force members; together with 24,000 public servants, they form the Government of Canada's "defence team," which is broadly tasked with the defence of Canada, the defence of North America, and with contributing to international peace and security, in that order (DND 2016). Like many militaries across the globe, CAF is broken down into three main branches that serve in distinct "environments": the Royal Canadian Air Force (RCAF), the Royal Canadian Navy (RCN), and the Canadian Army. Each branch contains its own set of trades, resources, and capabilities, as well as its own distinct cultures and historical traditions, many of which precede Canadian Confederation (Morton 2007).[24] The Army has the most personnel, as well as the most reservist regiments (Bercuson 2008). During most operations, regardless of type, the Army provides most of the "boots on the ground" while RCAF and RCN provide important support, such as materiel transport or air firepower (Granatstein 2002). Some missions, such as fighter jets targeting terrorist combatants in the Middle East or Her Majesty's Canadian Ships (HMCS) patrolling the Mediterranean, RCAF and RCN, respectively, play the main, "tip-of-the-spear" part of a mission.

While CAF doctrine follows "mission command," which means that once direction has been given, the lowest level officer in command is free to pursue their mission within given constraints, all three environments are ultimately overseen by Canadian Joint Operations Command (CJOC) in Canada's capital, Ottawa. The only parts of CAF not overseen by CJOC are the Canadian Special Operations Forces Command and the North American Aerospace Defense Command. The former is separate due to its command of highly classified, usually anti-terrorist missions, and the latter because it is a joint command with the United States. Domestic Joint Task Forces (JTFs) representing the six regions of the North, Pacific, West, Central, East, and Atlantic all see their chain of command link up into CJOC (DND and CAF n.d.a).

Each JTF is responsible for domestic operations in its region, and can pull resources from each branch, regular force, and reservist unit within its region to respond to requests for support. Under special circumstances a unit "owned" by one JTF can be placed under the command of another JTF should resource needs require it. JTFs of the West, Central, and East are all commanded by Army officers in command of the

Army division that corresponds to each of those regions, while the JTFs for the North, Pacific, and Atlantic are commanded by the Navy officers in command of the maritime forces that correspond to each coast (DND and CAF n.d.d).

Unlike the branch-specific formations within each region, JTFs are more a collection of all the military resources in a specific region than a unit with a specific identity. For example, soldiers from regular force infantry battalion A, Navy reservist regiment HMCS B, and a Hercules from RCAF base C may be called upon to support Joint Task Force X during a domestic disaster response, and overall coordination will occur at the JTF level, but each soldier will identify with and be commanded by their specific branch and unit.

As per Canada's national defence priorities outlined above, military units across all regions may be involved in non-combat domestic operations but need to be attuned to potential security threats to Canada. Throughout the 2010s, specific security concerns identified by the Government of Canada as key focus areas for the defence team were Russian aggression and threats of international terrorism (DND and CAF 2016). These concerns are defined by a novel blend of threats from non-state terrorist organizations and the re-emergence of state aggression. The former is epitomized by the grip extremist groups like the Islamic State of Iraq and the Levant (ISIL) hold on pockets of the Middle East, and the latter is characterized by Russia's aggressive action on Ukrainian soil (Stoltenberg 2015). Given that both of these types of threats are occurring simultaneously, the international security context is not understood as returning to either conventional state-on-state conflict situations or as a world defined by non-state actor violence. The twenty-first century is characterized as a hybrid of these two (Hoffman 2007; Deep 2015).

The recognition of "hybrid" security threats means that DND policy and CAF capability are orientated towards mitigating simultaneous state and non-state threats, including state threats that masquerade as non-state threats. While natural hazards due to climate change are included as serious threats in public consultation defence reviews (DND 2016), CAF capabilities are not tailored specifically towards natural disaster response in the way they are for the threats of Russian aggression (increased Arctic operations with drone surveillance) or international terrorism (continued development and deployment of the terrorism-fighting special forces unit, JTF 2).

The assessment of CAF effectiveness in any task requires in part an assessment of the financial resources that support CAF. Canada's defence spending is not the 2 per cent of GDP stipulated by NATO guidelines, and the country's relatively safe borders ensure that the use of public money on defence versus other priorities remains a perennial policy debate. However, twenty-first-century CAF operations at home and abroad demonstrate an organization with significant capacity to engage the development of perceived threats, be they directly human or natural. Since the end of Canada's 2001 mission in Afghanistan, which lasted over twelve years and saw over 40,000 troops deployed, CAF has remained engaged in an array of operations. Prominent operations abroad include Operation REASSURANCE (for which CAF has been expending considerable resources from all three of its branches to support NATO assurance measures against further Russian encroachment on Ukrainian soil after Russia's annexation of Crimea in 2014),[25] and Operation IMPACT (in which Canada's special forces have been performing a train and assist mission in northern Iraq in support of anti-terrorist engagements). While operations abroad gain significant media attention, many domestic operations are also active, and are especially important to understanding CAF capabilities and the context from which the CAF's disaster response emerges. An overview of some of these operations demonstrate the ongoing nature of CAF's work, which challenges the basic understanding of a military as "waiting in the barracks" to be called up for intermittent, discrete events. Prominent domestic operations include:[26]

1 Operation LENTUS, which establishes a plan to aid communities should municipal and provincial resources be depleted during a natural disaster. Op LENTUS pulls from all three CAF branches and has been enacted in an array of hazards across many provinces. Recent Op LENTUS deployments saw troops respond to a destructive New Brunswick ice storm in early 2017 and flooding in Quebec in the summers of 2017 and 2019. As Op LENTUS is seen by the Government of Canada as essential to keeping Canadians and their communities safe and is the operation title under which CAF responds to domestic natural disasters, it draws most of this book's attention.
2 Operation PALACHI is a part of an ongoing avalanche-control program in Rogers Pass in British Columbia that prevents avalanche blockage of essential infrastructure. This operation often continues until the spring of each year. Op PALACHI, while narrower and more specific in scope, is like Op LENTUS in that its mandate – keeping Canadian infrastructure safe from a natural hazard – does not include a potential human enemy combatant.

3 Operation LIMPID forms CAF's routine domestic surveillance presence across land, sea, air, space, and cyber areas. This ongoing Op is key to the early detection of threats to Canadian security. Op LIMPID is characterized as Canada taking direct responsibility for its own defence, in conjunction with – but without an overreliance on – the United States.
4 Operation NANOOK has since 2007 formed the largest CAF presence in the North and is seen as a cornerstone for maintaining Canada's sovereignty. Op NANOOK is intended to demonstrate that the Government of Canada prioritizes the defence of Canada's North as key among the defence team's main roles.
5 Operation NEVUS ensures the integrity of communication between Canadian Forces Station Alert and Ottawa. This annual maintenance Op allows an unbroken link between Canada's capital and its high arctic. Op NEVUS is intended to show that the domestic responsibilities of the defence team stretch beyond critical resource and population centres to every area of the country.

While the basic nature of Canada's military command structure does not change at home or abroad, domestic operations do differ from operations abroad at several levels. On the military side, specific domestic operations doctrine exists, which – through rooting itself in domestic law and military research on best practices specific to the type of domestic deployment – guides and constrains how CAF conducts itself on Canadian soil. This doctrine makes crucial distinctions such as "assistance to civilian authorities" versus "assistance to law enforcement," while also characterizing what effective civilian–military collaboration looks like according to military research.

On the civilian side, domestic operations do not carry the same potential for criticism as missions abroad. In Canada, there tends to be two objections that are routinely brought up when significant operations abroad are considered; one objection is economic while the other is about identity. Given Canada's relative geographical safety, and given that the US has an incentive to defend Canada to protect itself, the Government of Canada has historically been sympathetic to, and indeed at times made, economic arguments for spending less proportionally on defence than American administrations to achieve continental defence (Segal 2002). Indeed, historically only when American governments threaten to spend less on collective defence do Canadian leaders indicate a need for "hard power" and consider raising the defence budget (Global Affairs Canada 2017). Regarding the identity-based objection, the Government of Canada has often faced domestic pressure to differentiate the country from what is seen as the more aggressive elements of American foreign policy, such as in the case of the Iraq War and with regard to the continental ballistic missile defence system (Bercuson 2003; DND and CAF 2014). Neither the economic nor identity objections appear to affect domestic non-combat missions. While some provincial premiers may be reluctant to "call in the cavalry" for fear of appearing to have lost the battle themselves, and despite the small community here and there that prefers not to "see green" in their neighbourhoods, the Canadian public on the whole has generally reacted positively to Canadian soldiers deployed domestically to fight natural hazards.[27]

The caveat of natural hazards above is an important one given the history of CAF deploying domestically as an aid to law enforcement. Not only do many First Nations have negative views of CAF due to the military being deployed to diffuse Indigenous protests during the Oka Crisis (Smith 2017), but the military patrolling the streets of Montreal during the 1970 October Crisis has become a negative historical moment for many, and a clear sign of Canadian state oppression to sovereigntists, despite public opinion polls in and outside of Quebec heavily supporting the military presence at the time (Belanger 2000). Indeed, given the salience of those events it may be surprising that CAF is welcomed when natural hazards become disasters. Human behaviour, however, differs significantly depending on the type of threat that a community faces. Natural threats tend to see solidarity, generosity towards strangers, and acts of extreme selflessness, while events characterized by the use or threat of violence manifest the looting, rioting, and distrust of others so often portrayed in films and mainstream news reporting on disasters (Drabek 1986, 2010).

CAF employs the equivalent of more than a third of the entire federal public service (Government of Canada n.d.) and pursues complex tasks not found – or not legal – in the civilian sector. In the contemporary moment, CAF is situated in a hybrid security context with the mandate to – along with Canada's NATO allies – inhibit Russian aggression as well as diminish non-state terrorist organizations. Despite this security focus, and the training and equipment developed for combat, CAF runs several non-combat domestic operations and is increasingly called upon to respond to domestic natural disasters. The two, perhaps in tension roles of CAF – existing primarily as a war-fighting organization while simultaneously acting as a substantial humanitarian responder – need to be kept in mind when assessing how effectively CAF responds alongside civilian organizations during domestic natural disasters.

## Conclusion

Disaster response in Canada is a multi-governmental affair that is complicated by an institutional and historical gap between the federal and municipal governments, and by provinces that have substantial operational capacity compared to their counterparts in other federal states. CAF, an organization developed first and foremost to fight wars, is the one organization that links all three levels of government during domestic disaster response. The nature and quality of this link has implications not only for the academic literature, but for Canada's emergency management system as a whole.

# 2 Assessing Disaster Response through an Original Collaborative Framework

**Introduction**

The Canadian Armed Forces (CAF) plays a regular and substantial role in domestic disaster response, which presents the following puzzle: How effectively does CAF contribute to Canadian emergency management (EM), especially considering that CAF is not only the main federal organization that works directly with each level of government on the ground during the response phase of significant disasters, but is an organization developed primarily for the systematic and legal application of force? Scant attention has been paid to this puzzle, but relevant disciplines can guide its investigation. EM assesses disaster response effectiveness when different levels of government and multiple organizations respond together; civil–military relations assess the dynamics between armed forces and civilian governments when they work together; and public administration assesses the broader phenomena being analysed when studying governmental disaster response – the implementation of public policy.

While each discipline differs in focus area, all of them contain contemporary and influential niches that identify interorganizational collaboration as the crucial component to achieving desirable outcomes in disaster response, joint military–civilian projects, and public sector initiatives writ large. The puzzle of CAF effectiveness in Canadian EM can be addressed through describing, evaluating, and suggesting improvements to the nature of collaboration between CAF and the different levels of government during domestic disaster response.

**A New Framework for an Unaddressed Puzzle**

*Academia's Blind Spot: CAF as "Invisible" Pillar*

Almost no systematic analysis has been pursued where the role of CAF in EM is the primary object of study. Herspring's (2013) extensive research on Canadian civil–military relations, for example, barely mentions domestic disaster response. Wide ranging historical analyses of Canada's reservist regiments, specific battalions, one of the three branches, the entire CAF, or defence policy all elaborate on culture, traditions, individual leadership, battle performance, specific wars, technological development, or political debates in Ottawa to understand the nature of the military in Canada (Granatstein 1990, 2005, 2014, 2015, 2016; Bercuson 1999, 2008, 2015; Morton 2003, 2007), but the history and role of the military in domestic natural disaster response is an afterthought, if mentioned at all. Similarly, academic assessments of the contemporary CAF focus on conventional understandings of the armed forces as an organization mandated to take on combat or peacekeeping operations overseas (Ring 2009; Jones 2010; Warner 2013; Saideman 2016). While the impact that civilian domestic politics has on the nature of CAF is a common academic theme, along with assessments – and usually criticisms – of non-combat developments such as the heavy focus on peacekeeping in the 1990s and attempts at the "civilianization" of CAF in the preceding decades (Kasurak 1982; Herspring 2013), natural disaster response as a CAF function is rarely mentioned, never mind systematically addressed.

The academic analysis of CAF is also heavily situated within the disciplines of history and political science (and its sub-disciplines of international relations, security studies, and civil–military relations). Despite CAF being a large bureaucracy that pursues complex tasks not found – or not legal – in the private sector, and despite employing the equivalent of more than a third of the entire federal public service (Government of Canada n.d.), precious little has been written on CAF from the vantage point of public administration, where CAF as an implementer of public policy is assessed. A 2017 search of the *Canadian Public Administration* journal database for "Canadian Forces," "Canadian Armed Forces," and "military" yielded exactly one article that assessed the Canadian military as a bureaucratic organization; it is more than a few decades old and does not pertain to EM (see Kasurak 1982). Other articles in Canadian public administration that are loosely related to CAF focus on the civilian development of defence policy (Hartfiel 2010), not on the implementation of defence policy by the military. Only one scholar has assessed, albeit briefly in two articles, CAF as a bureaucratic organization involved in domestic disaster response, and his background is in journalism (Scanlon 1998, 2005).

The "CAF gap" in public administration and EM becomes even more pronounced when searches are made for EM military roles in comparable federal states such as the US and Australia.[1] The extensive database of the University of Delaware's Disaster Research Center boasts over 100 American case studies where the primary focus is on the impact the US military has on disaster responses (for example, Morris, Morris, and, Jones 2007).[2] The hazards in these studies run the full gamut from floods to wildfires to hurricanes to industrial accidents to violent events, and the military agencies responding range from the federal US Air Force to the US Coast Guard to a specific state's National Guard. Comparatively, studies where CAF is the focus yield exactly three articles in the database, all three authored (and in one case co-authored) by the same individual (Scanlon 1998, 2005; Scanlon, Steele, and Hunsberger 2012).[3] The population size difference between the two countries does not explain the discrepancy in the amount of academic analysis, especially considering that (1) other regularized elements of Canadian public administration and EM receive significant scholarly attention in Canada; and (2) Australia, a country of comparable size, produces significantly more studies on the role of their military in domestic EM.

CAF plays a regularized role in domestic EM and may provide the one consistent link between the federal and other levels of governments during disaster response, and yet almost no scholarly attention has been paid to the nature and implications of that role. With no existing findings, the disciplines of EM, civil–military relations, and public administration have to guide the way with data relevant to the puzzle at hand.

*Emergency Management: Interorganizational Collaboration as Effective Response*

Interorganizational competition generally has negative implications for disaster response while interorganizational collaboration generally has positive implications. This is not necessarily intuitive. Competition – incentivized by a market for disaster response – could theoretically allow a more effective disaster response system. Private companies contracted to governments or individuals to protect lives and property during hurricane season, for instance, may respond faster and with better technology than government coordinated emergency agencies. The need to outperform the competition – to win more contracts and customers – would mean that companies face little incentive to collaborate during a response through sharing information or resources, but could mean that each company pursuing an outcome of maximum lives and property protection – to increase the desirability of their brand and increase their market share – would allow an effective response at a system level.[4] The same could be true for public sector organizations;

lack of collaboration between the Red Cross and the Salvation Army may be good for disaster response since such lack of collaboration is indicative of competition for limited government contracts, which – as with the private companies – means that each non-profit will attempt to maximize its success compared to when it faces no competition and no organization against which to compare its outcomes (i.e., by providing only one contract that gives one non-profit a monopoly over EM services or by guaranteeing different non-profits contracts for different areas of EM).[5] Competition between government agencies, in turn, may also be intuitively hypothesized to be healthy for disaster response. The negative outcomes that may result from lack of information and resource sharing could be made up for by agencies that are forced to compete for government resources (i.e., an agency with a monopoly over disaster response may have less incentive to maximize the protection of lives and property compared to one whose future funding depends on how it compares to sister agencies).

The existing body of EM research, however, largely contradicts hypotheses that predict increased disaster response effectiveness when organizations compete rather than collaborate. While the literature is rife with examples of interorganizational competition, such competition is usually identified as a main culprit in disaster response *in*effectiveness. Indeed, even in its infancy, disaster research identified lack of collaboration among organizations, including between levels of government, as a central problem to effective disaster response (Rosow 1955; Williams 1956; Form and Nosow 1958).[6] Collaboration was characterized by these researchers as beginning with basic interorganizational coordination, such as the minimization of redundancies (e.g., if organization A is procuring resources to accomplish task X, then organization B will focus on task Y), but coordination ideally develops into collaborative processes such as the adoption of unified strategies, the ability to influence each other's actions, the sharing of resources, the streamlining of communication among different actors, and individuals from different organizations working together on the ground in a way that blurs lines between which organization is "in the lead" or "in control."

Again, there is nothing necessarily intuitive about such blurring of lines leading to effective disaster response. Clear divisions of labour, the specialization of labour, clear lines of organizational accountability, and a degree of hierarchy between organizations that allow one organization to call the shots amidst a crisis can reasonably be posited to be crucial to effective response. However, the chances of effective disaster response outcomes[7] – such as quickly delivering the right amount of physical and symbolic resources, swiftly getting individuals to safety, the protection of private property, the maintenance of essential public infrastructure, seeing to the psychosocial needs of individuals who experience the event, the maintenance of a community's values and beliefs, the ability for organizations to learn lessons from an event, and generally matching the supply of relevant expertise and capacity with corresponding on-the-ground demands – has been demonstrated since the 1950s to be improved by interorganizational collaboration.[8]

What specific components, then, of interorganizational collaboration can be assessed to determine response effectiveness during disasters? At a minimum, interorganizational collaboration requires *information sharing* among organizations. As no centralized agency can effectively produce an accurate flow of communication down the line to front-line organizations, those organizations need to be able to share information with each other.[9] If emergency response plans are not distributed, organizations are not allowed a representative in key Emergency Operations Centres, organizations are not allowed at key decision-making tables, communication is not maintained during response actions, new information is not distributed as the event unfolds, organizational strengths and vulnerabilities are not shared, and access to key individuals from specific organizations are not allowed, then response organizations will have to rely on the inherent limits of a central agency for information distribution and will therefore not have the capacity to work swiftly and directly together. Interorganizational collaboration, and hence effective disaster response, is not possible without high levels of information sharing across all responding organizations (Altay and Pal 2014; Altay and Labonte 2014; Olsson 2014).

Above and beyond information sharing, a disaster response organization that hopes to work in the absence of a limited central agency, and work together with other organizations on response actions to the point of organizational overlap, needs to be able to influence other organizations and be open to being influenced by them, albeit in a manner that does not seek control over the disaster response writ large (Kapucu and Garayev 2013; Carr and Jensen 2015; Zhang et al. 2016). Such *non-manipulative influence* entails decision-making capacity – not just presence – at key decision-making tables, the capability to change the actions of others not through threats but through persuasion, as well as a lack of institutional constraints that inhibit the ability to follow through on suggestions of other organizations. In a similar vein, organizations hoping to achieve interorganizational collaboration need to be *flexible* – willing to reach a decision not originally anticipated by an organization (Alison et al. 2015; Constantinides 2013) – and *supportive*, which means being open to the pooling of physical and non-physical resources to aid disaster response goals (Wells et al. 2013; Lee 2011; Stevenson et al. 2014).

As non-manipulative, flexible, and supportive as some organizations may be, conflicts still arise, especially in the intense environment of disaster response where each response action can potentially have an impact on saving human lives. A conflict resolution process is therefore required, albeit one that does not rely on a single powerful arbiter that decides on the "right way to go," as that would resurrect the problem of having a single agency in control of the disaster response. The conflict resolution process needs to be collective. Even under the time constraints of disaster response, the response system improves if conflicts are explicitly brought to a collective table where the input of other organizations is sought (Kapucu and Hu 2016). Collective conflict resolution requires a high degree of *trust* in the goodwill of other organizations, which is a component relevant more generally to disaster response organizations being able to work together without receiving specific directives from a single agency, and being able to work together on the ground. When organizations trust each other, they display a willingness to work with another organization over time, to have an individual from other organizations dictate actions (including the use of resources) for the organization, and to share data. Trust is essential to the sort of interorganizational collaboration that leads to effective response (Saab et al. 2013; Curnin et al. 2015).

The components of interorganizational collaboration that emerge from EM are information sharing, non-manipulative influence, flexibility, support, collective conflict resolution, and trust. The Canadian disaster response system can be assessed for these components, but first other relevant disciplines need to be assessed for how they can inform the question of how effectively CAF is integrated into that system.

### Civil–Military Relations: Shared Responsibility and CAF Doctrine

Civil–military relations is a multidisciplinary area – albeit made up mostly of political scientists – united by its object of study: the dynamics between a country's armed forces and its civilian government and/or society at large.[10] The focus is on how the arm of the state that most fundamentally operationalizes the state's monopoly on violence – to protect the state against internal and external threats – interacts with the state's civilian elements. Foundational texts on political strategy (e.g., Machiavelli's *The Prince*) and military strategy (e.g., Sun Tzu's *The Art of War*) are included in discussions of civil–military relations in so far as they describe how civilian authorities can employ armies to advance their goals, or, indeed, how armies can use civilian authorities to advance *their* goals (Freedman 2013).

While the discipline often assesses the role of militaries in authoritarian or totalitarian states, the focus here is on civil–military dynamics in a democratic context, where the military is under the legal control of civilian authorities that achieved their power through legitimate elections. This niche is found in the academy as well as in militaries themselves. Armed forces in advanced democratic countries have branches and/or chains of command that focus not only on weapons, combat, and training development but also on strategy development, which includes how most effectively to work with civilian organizations during domestic operations (Canadian Joint Operations Command 2015).

An overview of both the academic research on civil–military relations in democratic contexts as well as CAF's domestic operations doctrine, which is the main Canadian output of military research on the topic, demonstrates that the value EM places on interorganizational collaboration dovetails with the value contemporary civil–military relations place on a similar concept called "shared responsibility." As EM criticizes applying hierarchical or competitive organizational relationships to the disaster response context, so does civil–military relations criticize constructing the civilian–armed forces dynamic as one defined by a clear division of labour. Such a construction, popularized by Samuel Huntington, was the established model for democratic civil–military relations during most of the twentieth century following the Second World War (Herspring 2013). On the uses of a country's armed forces, the model conceptualized civilian authorities (i.e., politicians and bureaucrats) as policy-formulators and military personnel as implementers. Coordination was required even while collaboration was rejected. Constraining the formulation phase to civilians served the normative goal of democratic control, especially pertinent in the wake of the Second World War, while leaving implementation up to the military ensured that the perceived ineffectiveness and inefficiency caused by civilian authorities – driven as they are by political incentives or slowed down by bureaucratic procedure – was avoided.

However, the late twentieth century up to the contemporary moment has seen the Huntington model increasingly vulnerable to criticism. Proponents of shared responsibility note that a clear division of labour based on formulation versus implementation (a) does not reflect what sometimes occurs in reality, and (b) is not effective at achieving goals when it does occur in reality. Indeed, extensive comparative research on militaries from democratic countries around the world, including CAF, stress that civil–military relations that emphasize bidirectional collaboration between military and civilian leaders on both policy formulation and implementation allow for more effective outcomes on an array of operations and missions.[11]

Note that "implementation" here refers to the strategic, not the tactical, level; while shared responsibility sees a role for civilians in an operations centre, the concept does not suggest that civilians untrained in combat conduct a raid.[12] This is largely because the technical and physical skills required for that most military of functions – armed combat – is not something a non-soldier can be expected to do, or do well. However, the shared responsibility niche is as poor in case studies of the Canadian domestic response to natural disasters as the general civil–military relations literature, and so cannot make claims about non-soldiers with the technical and physical capacity to effectively perform on-the-ground jobs, as many civilians can during a natural disaster. Indeed, given shared responsibility's focus on collaboration during the implementation phase, the Canadian domestic response to natural hazards is an interesting test for a context that may benefit from military–civilian collaboration at a tactical level.[13]

While the civil–military relations niche that promotes collaboration between civilian and military leaders focuses largely on non-domestic combat contexts, CAF does elucidate its perspective on effective civil–military relations during domestic non-combat contexts through its domestic operations doctrine. This doctrine is mainly intended for military personnel as it is the actual code used to inform and constrain CAF actions taken during operations on Canadian soil. As such, the doctrine functions simultaneously as a best practice guide and a semi-legal framework. While CAF's domestic operations doctrine does not explicitly use academic phrases like "shared responsibility," the doctrine's prescriptions dovetail with the shared responsibility approach. The doctrine stresses that multi-organizational integration – defined as achieving a common purpose, creating a common situational understanding, establishing an interoperable planning process, maintaining a unity of effort with personnel from different organizations participating in overlapping activities, and discerning common measures of effectiveness – "delivers the most effective and efficient results ... [Early] and continuous interaction [between military and non-military] organizations is essential" (Canadian Joint Operations Command 2015, 3).

CAF's domestic doctrine is therefore clearly in line with the concept of shared responsibility: Common measures of effectiveness put CAF in a policy formulation role, while allowing personnel from different organizations to participate in another organization's activities puts civilians in an implementation role. Furthermore, the doctrine explicitly states that the effectiveness of domestic operations is increased when military resources are provided to support the objectives of non-military groups, when partner response organizations are allowed to participate in some aspects of CAF training, and when CAF members are free to participate in some aspects of partner response organizations' training (Canadian Joint Operations Command 2015, 3).

Both CAF and proponents of shared responsibility see civilian and military personnel working together on formulating response actions as well as working together on the ground implementing such response actions.[14] While CAF personnel will be in uniform, and individuals with specific skills will tackle corresponding tasks, belonging to a particular organization does not on its own preclude individuals from working with other organizations or planning versus performing response actions.

What specific components, then, allow such shared responsibility to occur? At minimum, and like the interorganizational collaboration prescribed by EM, shared responsibility requires information sharing. Furthermore, all the components identified and described as important to interorganizational collaboration in general are important to shared responsibility between the military and civilian organizations. Both the military and the civilian side need to be able to influence each other in non-manipulative ways, be flexible and supportive, and solve conflicts in a collective manner if they hope to share responsibility during disaster response. CAF domestic operations doctrine does, however, introduce extra components that need to be included when a disaster response system that includes an armed force as a main player is being assessed. Given that (a) the military's operating procedures are rooted in linking discrete, specific actions to discrete, specific outcomes, and that (b) certain operational criteria must be achieved for the military to maintain the organizational capacity that makes it useful to disaster response in the first place (i.e., speedily marshalling hundreds of individuals to complete an array of technical and physical tasks), two additional components can be discerned from CAF's domestic operations doctrine to complement those components already identified as important to EM's interorganizational collaboration.

The first is *coordination*, which, unlike the "higher level" collaborative elements of non-manipulative influence and support that see individuals from different organizations working together in a way that blurs organizational boundaries, requires different organizations to first and foremost agree on some practical elements of the domestic response – namely, a common vision of what needs to happen for the response to be over (i.e., to reach the "end state"), a general agreement on what actions to take to reach the end state, a mutual understanding that having more than one main organization involved in the response is more effective and efficient, an appreciation for the impact that one organization's activities will have on the other response organizations, and consultation – whenever feasible – before taking response actions (Canadian Joint Operations Command 2015, 3).

The second is *integration*, which adds to the components of information sharing, non-manipulative influence, support, flexibility, and collective conflict resolution by emphasizing the operational characteristics that would allow such components to occur. Integration requires that CAF and civilian organizations not just agree on a common end state, but that each organization shares a common purpose in delivering that end state. CAF and the other response organizations need to have the same understanding of the scope and nature of the event (i.e., the same "situational outlook"); plan together as well as produce plans that can be used by all response organizations; and allow personnel from one organization to work in another organization.

The Canadian disaster response system can be assessed for all the above components, but how does public administration inform such an assessment?

## Public Administration: Beyond Silos

Public administration is a multidisciplinary field that includes the study of management in the public sector; the formulation and implementation of public policy; the economic, sociological, and political components of the public policy process writ large; and the governance of the state itself, all of which means that the field – like EM, and civil–military relations when military research is included – is guided by both scholars and practitioners (Hodgetts 1997). The fact that neither group dominates the field, the breadth of activities that involve governments, along with varying and competing definitions on what the "public sector" entails, ensures that many competing definitions of public administration exist. Definitions may be state-focused, include non-profits or private companies that deliver government programs, or even umbrella any groups from civil society that act in the public sphere (Frederickson 1991, 2012).

Within academia, public administration can include a focus on policy analysis, where the pedagogical outcomes are skills taught to serve a particular client in the public policy process, or a meta-assessment of a jurisdiction's policy process itself, with no client but knowledge itself in mind; both these "streams" are informed by political scientists, economists, sociologists, legal scholars, as well as academics who simply identify as public policy and/or administrations scholars (Botha, Geva-May, and Maslove 2017).[15] Given the focus here on how effectively CAF is integrated into the domestic disaster *response* system, the public administration areas of most interest are those that seek to understand the management of government organizations involved not in the formulation of public policy, but in the implementation of public policy.[16]

The public *policy* of EM in Canada, which is anchored in municipal, provincial, and federal EM frameworks and legislation, is for government organizations to be the main players in disaster response, from the management of the response to the provision of resources.[17] An assessment of EM policy would focus on how, for example, the non-profit or private sector could be incentivized to be more involved to achieve cost reduction or effectiveness outcomes. An assessment of EM policy *implementation*, however, requires a focus on those organizations that anchor Canadian disaster response as it currently stands, which means each level of government, their emergency management organizations (EMOs), and – if the disaster is large enough – CAF, are in the spotlight. This implementation focus means that an array of otherwise beneficial and interesting public administration scholarship, such as the role of non-profits and the private sector in achieving public policy goals or the role of perverse incentives in frustrating public policy goals, are largely beyond the scope of this book. An economist's game theory matrix may explain why the rules established by certain municipal bylaws and tax incentives lead to continuous development on flood plains that cost taxpayers in the long run, or a feminist critique of disaster recovery programs may illuminate the gendered forms that post-disaster compensation may take. Such disciplines and methodologies, however, are not the most directly relevant to a research puzzle orientated towards the dynamics between CAF and civilian government organizations – the main players in the implementation of EM policy as it currently stands – during the disaster response phase. Addressing this puzzle requires those disciplines and methodologies within public administration that assess the management of public policy implementation.

Public management contains the relatively recent approach of new public governance (NPG), which – like interorganizational collaboration from EM and shared responsibility from civil–military relations – stresses the importance of relationships between public sector organizations that are collaborative in nature and not defined by strict boundaries, hierarchy, and/or competition along organizational lines (Huxham et al. 2006; Mandell and Steelman 2003; Leach 2006; O'Leary and Bingham 2009). This approach, while general, is particularly relevant to disaster response, as NPG – sometimes called "collaborative governance" by public administration scholars working within EM – prescribes management practices whereby leaders work together across hierarchical, organizational, and even jurisdictional lines when confronted by complex and unpredictable phenomena that are not easily perceived in the same way across stakeholders (Agranoff and Mcguire 2003; O'Leary and Bingham 2009; Simpson 2011) – a characterization that fits a disaster event.

The practices prescribed by NPG dovetail with the components of effective disaster response discerned from the other two disciplines, such as the development of unified strategies (i.e., non-manipulative influence), formal procedures for incorporating stakeholders (i.e., integration), and processes for consensus-building (i.e., collective conflict resolution) (Agranoff and McGuire 2001, 2003; Vigoda and Gilboa 2002; Ansell and Gash 2007). Just as EM's interorganizational collaboration approach rejects one organization maintaining monopoly over – or many organizations competing during – disaster response, and the civil–military relations' shared responsibility approach rejects constraining military versus civilian organizations to specific tasks, the consensus-building element inherent in NPG likewise requires a high degree of collegial communication between public sector organizations and rejects forms of "siloization" between departments and agencies.[18]

Case studies across various policy areas – from environmental policy to military policy to general policy-making at a municipal level – demonstrate that policy goals in the context of complex challenges that require input from different levels of government and types of organizations are more likely to see success if the public sector organizations involved formed a "collaborative governance regime" (Emerson and Nabatchi 2015). Organizational relationships within these regimes display all the components thus far discerned from the other two disciplines, as well as two extra components. While organizations within collaborative governance regimes practise information sharing, non-manipulative influence, flexibility, support, collective conflict resolution, coordination, integration, and trust, they also value *satisfaction* and processes to deal with *conceptual difference*. The likelihood that public sector organizations will continue to work together, and do so collaboratively, is increased when organizations express approval of other organizations' actions, hence the importance of satisfaction (Bhakta Bhandari, Owen, and Brooks 2014; Schmeltz et al. 2013), and organizations cannot effectively define problems, plan solutions, or resolve conflicts together if they have fundamentally different perceptions of the challenge at hand, hence the importance of identifying and dealing with conceptual.[19]

PUBLIC ADMINISTRATION: AN ECONOMIST'S LENS

The components of collaboration drawn from NPG and the other approaches are all "positive" in the sense that they describe a desirable (inter)organizational attribute. This is to be expected as in each discipline the components are discerned from approaches that aim to achieve specific, desirable outcomes: for EM's interorganizational collaboration, desirable disaster preparedness, mitigation, response, and recovery outcomes; for civil–military relations' shared responsibility, the successful formulation and implementation of defence/military policy; and for public administration's NPG, the completion of policy goals in the face of complex policy changes. For all these approaches, the disaster event or disaster process can be seen as the problem and the central object of study. For EM, disasters are the adverse event to be managed; for shared responsibility, disasters are the enemy, an operation's raison d'être, the dissolution of which would indicate the operation's end; and for NPG, disasters are the wicked policy problems to be addressed. In each case, the organizations and organizational attributes that successfully manage, defeat, or address disasters are theorized and/or studied. Similarly, CAF's domestic operations doctrine provides a practical framework that is meant to spark and guide effective operations on Canadian soil. Like the above approaches, the CAF doctrine does not elaborate at length on what may actively work against collaboration.

However, a framework that seeks to assess the effectiveness of Canadian disaster response cannot rely only on components that studies have demonstrated to be "important to effectiveness," even if a wide and varied body of literature is used. If the main disciplines relevant to this study's research question generally conclude that collaboration is crucial to effective disaster response, then a framework used to assess disaster response also requires components that indicate barriers to collaboration. Yet the approaches already discussed do not elaborate at length on significant barriers to collaboration in the context of joint military–civilian organizations responding to domestic natural disasters. An economic theory of the bureau from within public administration,[20] however, does explicitly theorize components that would inhibit interorganizational collaboration within the public sector. Indeed, in such a theory, *anti-collaboration* is pursued if it benefits individuals within a specific organization (Breton and Wintrobe 2008).

An economic theory of government organizations assumes public managers pursue the power and growth of their organizations above all else, which allows the theory to describe contexts where organizational anti-collaboration is logical, functional, and unexceptional in abstract terms. For example, if withholding information from other organizations could help organization X use that information to receive credit – and therefore prestige, influence, and funding – then, according to an economic theory of the bureau, organization X would withhold such information even to the detriment of the overall disaster response.[21] Organizations that operate according to an economic theory of the bureau would go even further; interorganizational trust would be treated merely as a means-to-an-end currency – one that would be exploited should enough power and prestige be on the line. *Empire-building*, *manipulation*, and *distrust* are therefore all signs that barriers to collaboration are in place and that individuals within specific organizations are pursuing their own self-interests (Olson 1965; Breton and Wintrobe 1986, 2008; Breton 1998).[22] These components are hinted at in EM, NPG, and civil–military relations, but only as departures from the ideal and not as dynamics to be expected whenever organizations pursue the same task – and certainly not as dynamics that could improve domestic disaster response.

There is one other component that undermines collaboration, but that does not fit into economic theories of the bureau as it is not sparked by organizational competition. EM and CAF's domestic operations doctrine describe contexts where interorganizational collaboration is inhibited simply due to a discrepancy of resources among organizations (Canadian Joint Operations Command 2015; Walker 1992; Miao, Banister, and Tang 2013; Stevenson et al. 2014; Kapucu and Garayev 2014; Ansell and Gash 2007; Raikes and McBean 2016). For example, disaster response organization X may not be sharing its information with organization Y because the latter does not have the technical expertise to interpret X's geographical information systems map, or because it may have the expertise but not the funds to procure a digital platform to carry the map. When the help of the military is requested specifically because of its unique physical resource capacity, then civilian organizations likely do not have the equipment to join some military tasks, such as transporting thousands of tons of materiel through airlift or building bridges in vulnerable soil conditions. In other words, *benign incapacity* may occur when some organizations do not have the technical, financial, or physical resources to collaborate with other organizations during the disaster response.

## *A Collaborative Framework*

Now that the three disciplines most relevant to the puzzle at hand have been assessed for approaches that will lead to effective joint

Table 2. A Collaborative Framework

| Research Puzzle Questions | Concept | Components |
|---|---|---|
| 1. What is the role of CAF in contemporary Canadian disaster response? [descriptive] | The *Presence* of Interorganizational Collaboration | - Information sharing<br>- Non-manipulative influence<br>- Flexibility<br>- Support<br>- Collective conflict resolution |
| 2. How effective is the civilian–military relationship during disaster response? [evaluative] | The *Quality* of Interorganizational Collaboration | - Coordination<br>- Integration<br>- Satisfaction<br>- Trust |
| 3. In what ways might the military contribution to Canadian disaster response be improved? [normative] | *Barriers* to Interorganizational Collaboration | - Empire-building<br>- Manipulation<br>- Distrust<br>- Benign incapacity<br>- Conceptual difference |

civilian–military domestic disaster response, the components that emerged from each approach can be used to inform a single framework that allows for the description, evaluation, and suggestions of improvement to the Canadian disaster response system (see Table 2).[23]

The first systematic assessment of how CAF is integrated into Canada's disaster response system could not afford disciplinary tunnel vision. EM, public administration, and civil–military relations, including CAF's domestic operations doctrine, need to inform how CAF's role is understood. New assessments often require new frameworks, which is the case here. The original collaborative framework built specifically to address this research puzzle opens up the analysis of specific disaster events to account for descriptive, evaluative, and normative results. How and to which events this framework is applied is the focus of the next section.

## Two Floods, a Wildfire, and a Hurricane

### The Impetus for Case Studies

The justification for prioritizing the public management literature – as opposed to policy-focused literature – in the theoretical framework was rooted in this book being oriented towards the *implementation* of EM policy on the ground. The justification for employing qualitative methods – case studies of responses to disaster events – also depends on this orientation. The puzzle at hand, and its attending research question, lies in the nature of intergovernmental and civil–military relations as disaster response operations unfold and does not assume a default correlation between formal policy and actual behaviour. Assessments of large quantitative datasets and/or applications of economic models may result in "ideal EM policy X in context Y," yet they will have to come with the usual *ceteris paribus* caveat and will say nothing directly about that variable essential to effective EM identified above – interorganizational collaboration during the response phase.

No quantitative figure – whether total (un)insured costs, levels of property damage, government expenditures, and so on – will directly indicate levels of collaboration during the response phase because the varying impact of different disasters, which depends – among other things – on hazard type, community size, community resiliency, the local economy, and local assets, may allow skyrocketing figures with high levels of collaboration (in which case the costs could have been even worse) or low figures with low levels of collaboration (in the case of a particular hazard missing expensive infrastructure). While the disciplines assessed above demonstrate a correlation between desirable disaster outcomes and levels of collaboration, methodological robustness warrants directly assessing the variable of interest over another variable with which it may be correlated. A qualitative approach that directly assesses emergency managers and leaders across levels of government and the military for levels of collaboration during disaster response is therefore the method employed here.

There are practical and methodological reasons for the use of case studies as the specific qualitative approach employed. While EM phases are an ongoing cycle for the EM system, with mitigation in particular receiving continuous attention, the response phase itself is a discrete event that begins when a hazard initiates a disaster event and ends when the community undergoing the disaster can begin the recovery process. Disasters are, by definition, events that occur outside the routine adverse events experienced by a community, limiting the ability to identify enough similar events to qualify for a large-N study. This latter point is even more acute for disasters that require a significant CAF contribution, since such events are the largest non-routine adverse events faced by the country (anything larger would be a catastrophe that transcends the EM realm) and so are even smaller in number than disasters where CAF resources are not requested.

The nature of significant disaster responses as discrete, varying, and relatively small in number means that modern social scientific inquiries into disaster responses have relied heavily – and, until the late twentieth century, almost exclusively – on case studies (Perry and Quarantelli 2005). Indeed, the research credited to be the first social scientific assessment of disaster response is the case study of the infamous 1917 Halifax explosion, conducted by Canadian scholar and priest Samuel Henry Prince for his doctoral thesis at Columbia University (Prince 1920; Scanlon 1988).[24] While case studies face the well-established limitations of not allowing statistically satisfactory generalizability, they can be powerful explanatory tools for research questions that require qualitative analysis, especially when a comparative element is introduced to avoid mere description (Barton 1969; Perry and Quarantelli 2005), as is the case in this book. Indeed, this book veers away from the heavy use of single case studies in the Canadian EM literature that focus on only one jurisdiction, hazard, or event (Hale 2013; Catto and Tomblin 2013; Scanlon 2012; Henstra 2011).

Case studies of specific disaster events also allow emergency managers and leaders from different organizations involved in the same event to be asked about that event and its specific disaster response actions. Not only do more perspectives on a specific event/response action minimize the bias of using only one perspective, but (dis)agreement on events/response actions across levels of government and the military is itself an indicator of the components that define the presence and quality of, and barriers to, collaboration.[25]

*Case Selection*

The logic of explanation can be violated when selecting case studies on the dependent variable (Geddes 2003). Care was therefore taken to avoid such an outcome by selecting cases where (a) the dependent variable – the presence and quality of, and barriers to, interorganizational collaboration across levels of government and the military during disaster response – is unknown; and (b) the jurisdictional contexts and hazard types differ enough to minimize the likelihood that the selection process favoured cases that oriented around the dependent variable.

The selected cases build on the growing methodological practise of comparing subnational units to understand policy implementation. All cases are Canadian due to the orientation of the research puzzle, but also because subnational, cross-jurisdictional comparisons alleviate some of the typical limitations of small-N research designs, such as too many variables of difference when comparing across nation-states (Snyder 2001). The selected cases fit into the "diverse" method of case selection because – while similar enough to compare – the cases demonstrate a diverse range across the independent variables (jurisdictional context and hazard type) to explore the nature of the dependent variable (Seawright and Gerring 2008). To analyse the research puzzle and its attendant three questions as laid out in the collaborative framework (see Table 2), the selected cases needed to satisfy the following five basic criteria:

1 "Non-routine" emergencies: Hurricanes, floods, ice storms, and wildfires are the most common hazards in Canada that fit the definition of events that fall under the purview of EM policy and administration.[26] They are generally anticipated, but their exact timing, location, and impact cannot be predicted. They are not routine emergencies, such as those that are addressed by the existing emergency medical, firefighter, and police services. Neither are they so catastrophic that powerful government agencies in concert with private and non-profit actors cannot significantly reduce their impact.
2 Recent: EM policy and practice, including the formal Request for Assistance (RFA) process needed to officially initiate a CAF response, has developed significantly throughout the twentieth century and into the current one.[27] In particular, the first decade of the twenty-first century saw the practice of EM become a professional, accredited field; the RFA process became standardized across Joint Task Forces (JTFs) and hazard types; provincial EMOs grew into organizations of considerable scope and influence; and the entire federal EM framework, including national EM legislation, revamped in 2007. Selecting events before 2007 would therefore illuminate the ebbs and flows of EM over time, and so have historical significance, but such cases would not paint an accurate picture of what EM looks like in contemporary Canada.
3 Hazard type: Different hazards are required to hedge against the possibility that intergovernmental and civil–military relations are a function of institutionalized collaborative practices formed around only one type of hazard. For example, a potential finding of highly collaborative results between CAF and the Government of Manitoba may occur because the relevant JTF and that provincial government have significant historical experience managing floods, which limits any general claims about levels of CAF–civilian collaboration. The inclusion of case studies where other hazards led to disaster helps to mitigate such a limitation.
4 Jurisdictional location: In a similar vein, different jurisdictional contexts are required to balance against the possibility that intergovernmental collaboration has developed in one particular jurisdictional context. Intergovernmental dynamics and civil–military relations

may differ across provinces, and the generality of the research's empirical claims – already limited by the case study approach – will be inhibited through a less diverse sample. For example, if CAF–civilian collaboration is found in each case, despite their jurisdictional differences, then a more general claim about civil–military relations during disaster response can be made than if collaboration occurred in only Alberta.

5  Significant military contribution: A significant CAF contribution is required in each case study in order to assess the level of collaboration between CAF and civilian response organizations. Too small a contribution minimizes the opportunity for the type of response challenges that require collaboration in the first place. Significant CAF contribution introduces the possibility of coordination problems, the opportunity for empire-building, the ability to influence the response outcome, the possibility that a larger number of decision-makers may disagree conceptually on the nature of the threat, more organizational relationships that may or may not be trusting or satisfactory, and so on. In other words, the components that define levels of interorganizational collaboration between CAF and civilian organizations can only be assessed during disasters that warrant a significant CAF contribution. High or low levels of collaboration discerned from a response where a mere platoon's worth of soldiers helped a community would not paint a convincing picture of what Canada's disaster response looks like at a system level.[28]

There are four Canadian disasters that fit the above criteria: (1) Hurricane Igor, which hit Newfoundland and Labrador in 2010;[29] (2) the 2011 Assiniboine River flood in Manitoba; (3) the 2013 Alberta multi-river flood; and (4) the 2015 Saskatchewan wildfires.[30]

*Overview of Disaster Events*

An overview of each disaster event assessed is provided below in chronological order. Disaster events are broken into (1) the hazard, location, impact, and duration details that defined the disaster event; and (2) the municipal, provincial, federal, and CAF actions that characterized the response.

These overviews help build a picture of the events based on extensive archival research. The goal of such research was to identify cases that met methodological requirements, inform the preparation for interviewing civilian officials and military officers involved in the responses, and corroborate what such officials and officers claim about specific events and response actions.[31]

2010 HURRICANE IGOR | NEWFOUNDLAND AND LABRADOR
Event Details:

- Hazard/Location: Hurricane Igor was the strongest tropical cyclone of an already busy 2010 hurricane season. On 6 September 2010, a tropical wave accompanied by a broad area of low pressure developed near the Cape Verde islands off the western coast of Africa. By 12 September, the cyclone had reached hurricane strength as it moved towards the Americas. On 20 September, Igor had weakened and passed by Bermuda. However, forecasts underestimated Igor's ability to grow as it accelerated towards the Canadian Maritime provinces.[32] Tropical storm–force winds increased to 750 nmi wide and speed strengthened to 75 kt as Igor hit near Cape Race, Newfoundland, at 15:00 UTC on 21 September. The hurricane centre passed over the Avalon Peninsula's east coast for two hours after landfall. In terms of gale diameter (1,480 km), Igor was the largest hurricane ever recorded in the Atlantic Basin.
- Impact: The main mechanisms of hazardous impact were hurricane-force winds and rain-induced floods. Winds of up to 140 km/h battered infrastructure and inhibited mobility, while 150 communities became isolated due to roads and bridges washed away by rising rivers. A section of the Trans-Canada Highway in Newfoundland was washed away, leaving a ravine 30 m across. 70,000 hydro customers lost power while basements were filled with water. Entire homes were destroyed. Insurable claims were $65 million while non-insured costs exceeded $120 million. Approximately 5,000 trees were felled. While only one person died directly due to Igor, the difficult-to-measure psychosocial impact on individuals and communities appears to have been substantial given the unexpected extent of the damage.[33]
- Duration: Igor hit Newfoundland on the afternoon of 21 September and had moved on by 22 September. However, bad weather from Igor's "tail" continued for several days as the system moved northwest between Labrador and Greenland.

Response Details:

- Municipalities: Mayors quickly made public statements that their community's ability to respond had been overwhelmed. Thirty municipalities issued official state of emergency declarations, and the City of St. John's activated its emergency response plan. Even as provincial and federal resources became necessary, municipal officials remained a part of coordinating response and relief efforts for Igor's duration and aftermath.

- Province: A state of emergency was declared for eastern Newfoundland after communication with impacted municipalities. Provincial crews filled in roadway craters, cleared debris, and restored power lines *during* Igor. On 24 September, after the storm had abated, Newfoundland and Labrador requested military aid to supplement civilian power. Provincial crews and hired contractors were sent to complement military efforts. An EMO was set up to field ongoing requests for help. The province organized two ferries to transport supplies to isolated communities. The premier helicoptered into isolated communities to survey damage and initiated a housing strategy to accommodate those who had lost their homes to the storm.
- Federal Government: Department of National Defence (DND) personnel and military assets were prepared to deploy on 21 September in anticipation of provincial requests for aid. Public Safety Canada notified other federal departments on 23 September that provincial resources would not be sufficient. CAF was deployed on 24 September. The Canada Revenue Agency stated that those affected by Igor might be eligible for tax relief. Federal relief funds were made available through Disaster Financial Assistance Arrangements and the Joint Emergency Preparedness Program. The prime minister characterized Igor's destruction as the worst he had ever seen.
- CAF: Operation LAMA involved more than 1,000 regular and reserve personnel from all three branches of CAF. Over forty communities were assisted. Tasks completed included the construction of three temporary bridges, surveys and culvert checks of over 900 km of highway, signage construction, the transportation of goods, services and people (by air, land, and sea), and the production and distribution of drinkable water.
    - Duration: 24 September 2010 – 6 October 2010. 13 days.
    - Assets Deployed: the 4 Engineer Support Regiment (from Gagetown, New Brunswick); 200 reservists from Newfoundland and Labrador that formed a Domestic Response Company; HMCS *St. John's*, *Fredericton*, and *Montreal*; 3 CH-124 Sea King helicopters (12 Wing Shearwater); the CC-177 Globemaster III; 1 CP-140 Aurora aircraft.

## 2011 ASSINIBOINE RIVER FLOOD | MANITOBA
Event Details:

- Hazard/Location: Heavy precipitation in western Manitoba and Saskatchewan in fall 2010 and spring 2011 set the stage for the 2011 Assiniboine River flood. Swiftly rising spring temperatures saw 60 per cent higher build-up in the Assiniboine than the previously recorded highest peak in 1923. In 2011, the "epic" spring melting sparked an over 1,000 m$^3$/s peak, which initiated a 1 in 300-year flood. The former village of St. Lazare was hit first before flooding affected municipalities across the province, including Manitoba's second biggest city, Brandon.
- Impact: The main mechanism of hazardous impact was high water levels that saw heavy flow into homes, across essential infrastructure such as roads and bridges, and onto primary economic resources such as farmland. With 850 road closures, including sections of the Trans-Canada, seventy local states of emergency were declared. A total of 7,100 Manitobans were displaced, including 1,200 people from the City of Brandon. The year's end arrived with almost 3,000 people still evacuated. One Winnipeg fatality was connected to fast-flowing Assiniboine water. Property damage, agricultural losses, and victim compensation costs surpassed $1 billion. Note that the "natural" impact of the hazard was complicated due to the use of intentional flooding to minimize total assets flooded. Purposeful (i.e., human-initiated) water spill from the Assiniboine executed through breaching and water diversions were expected to cover 185 km and flood up to 150 homes.[34]
- Duration: Exact dates of hazard duration are difficult to establish due to the slow onset and decrease of spring melting-induced flood waters on the Canadian prairies. The flood began to seriously affect Manitoban communities in mid- to late April and continued having adverse impacts until mid- to late June.

Response Details:

- Municipalities: Many municipalities prepared for rising floodwater by constructing dikes. The extent of the flood, however, overwhelmed these measures. By early May, states of emergency were being declared by a variety of municipalities, including the City of Brandon.[35] Affected municipalities closed schools, cancelled local events, posted reminders to tourists to double-check hotel availability, and the worst hit issued evacuation notices.[36] Some rural municipalities had full- or part-time emergency managers who began coordinating a response while others allocated the EM function to a point person. With state of emergency declarations came calls to the province for response and financial aid. Some municipalities had a say in how breaches and diversions were to be used. However, ultimate decision-making power and response coordination rested with the province.

- Province: Manitoba Infrastructure's Hydrologic Forecast Centre (HFC) provided flood forecasts. The same department's EMO maintained overall control of coordinating response efforts. The province remained responsible for the many roads under provincial maintenance. As the flood grew worse, the province maintained its data-provision and coordination role while providing some crews to help overwhelmed municipalities. The Manitoba Emergency Coordination Centre (MECC) was set up by the EMO and ran for 103 days (compared to 33 days during the "Flood of the Century" in 1997). EMO leadership worked with regional EMO officers to assess and provide for local needs. The Health Emergency Management's Office of Disaster Management did the same with its regional officers, their focus being on the continued provision of health care. As the EMO and other agencies coordinated a response, water management experts in conjunction with political leadership began making decisions on where to breach and divert water to minimize overall property damage. Water was drained from Lake St. Martin and Lake Manitoba into Lake Winnipeg. A provincial state of emergency was declared on 9 May. A request for federal assistance, including help from CAF, had been made the evening before on 8 May. While the "whole of government" concept was not emphasized as it was during Alberta's 2013 floods, the province was the dominant actor in responding to the flood.
- Federal Government: Aboriginal Affairs and Northern Development Canada (AANDC) coordinated establishing dikes for First Nations communities.[37] An expanded financial program was established to supply engineering expertise and heavy equipment to First Nations. The Disaster Financial Assistance Arrangement was initiated. Over half a billion dollars of federal reimbursement was expected.[38] The prime minister visited the hardest hit flood zones, and CAF was deployed on 9 May.
- CAF: Operation LUSTRE involved more than 1,800 regular and reserve force personnel from all three branches of CAF. While more than 160 private residences were protected from flooding, the exact number of communities assisted is not clear.[39] Tasks completed

included repairs to dikes, monitoring of the dikes (from both land and air), assistance with voluntary evacuation, producing more than 167,000 sandbags, and laying half of the 891,000 sandbags used in the fight against the flood.
- Duration: 9 May 2011 – 26 May 2011. 17 days.
- Assets Deployed: 130 Navy personnel (mostly reservists from across the country); 70 Air Force personnel; 200 Army Reserve personnel (38 Canadian Brigade Group); 1,400 Regular Army personnel (1 Canadian Mechanized Brigade Group and 1 Area Support Group from Canadian Forces Base [CFB] Shilo and CFB Edmonton); 6 CH-146 Griffon helicopters (408 "Goose" from Edmonton and 400 "City of Toronto" from Borden, ON); 1 CH-146 Search and Rescue Griffon helicopter (424 "Tiger" from 8 Wing Trenton, ON); 1 CP-140 Aurora maritime patrol aircraft (407 "Demon" from 19 Wing Comox, BC).

## 2013 MULTI-RIVER FLOOD | ALBERTA
Event Details:

- Hazard/Location: Uncharacteristically heavy rain poured into southern Alberta in mid-June due to a high-pressure system "holding down" a low-pressure system in the south. More than 200 mm of rain fell in less than two days across regions southwest of Calgary. In Canmore, half the annual average rainfall fell in thirty-six hours while High River saw 325 mm in forty-eight hours. A "devil's cocktail" of heavy snow loads in the Rocky Mountains, steep watersheds, and saturated earth swelled the Bow, Elbow, Highwood, Red Deer, Sheep, Little Bow, and South Saskatchewan rivers and their tributaries to the point of heavy overflow.
- Impact: The main mechanisms of hazardous impact were high and strong water flows that damaged infrastructure, destroyed homes or rendered them temporarily dangerous, and inhibited public and private transportation. Up to 100,000 Albertans faced evacuation orders while five deaths were directly attributed to the floods. Calgary – one of the country's economic hubs and the province's largest city – experienced crippled infrastructure and issued evacuation orders that affected 75,000 people; the city's downtown core was largely shut down.[40] The floods' insurable damages were $2 billion, making it the costliest natural disaster in Canadian history at the time. Total damages exceeded $6 billion.
- Duration: 19 June 2013 – 12 July 2013. 23 days.

Response Details:

- Municipalities: The City of Calgary issued mandatory evacuation orders on 20 and 21 June. Officials dubbed 21 June a "family day" and asked people to stay home. Schools were closed. The mayor was particularly active providing updates on social media, as were some city councillors. The total population of High River – 13,000 residents – faced and followed evacuation orders that lasted weeks while other communities saw their residents return home.[41] Numerous other communities, including a number of First Nations reserves, likewise declared states of emergency on 20 and 21 June. The City of Medicine Hat evacuated 10,000 people before the floods hit.
- Province: The largely lauded provincial task force model established during the 2011 Slave Lake fire was put into effect. As other provincial actors focused on response, the task force began constructing a recovery framework during the response phase. The province emphasized a "whole of government" approach that saw lines of communication maintained between provincial agencies as they managed different aspects of the flood.[42] The Alberta Emergency Management Agency aided the overall response coordination of smaller communities while Calgary's Emergency Management Agency affirmed Alberta's substantially decentralized municipal system by largely managing that city's response. The Provincial Operations Centre (POC) was activated at its highest level for almost a month. The premier toured affected regions and promised that the province would play a role in financial recovery. On 20 June the request process for CAF help was initiated and by 24 June the provincial Treasury Board had approved the first set of funds to buttress recovery efforts. Uninsured losses incurred by municipalities and individual homeowners were to be covered by the province. On 28 June, Alberta declared its first ever provincial state of emergency as flood impacts continued to batter the town of High River.
- Federal Government: On 21 June, the prime minister joined Premier Redford and Mayor Nenshi to tour affected areas. CAF was deployed on the same day. The Disaster Financial Assistance Arrangements program was kick-started as the extent of the damage became clear. While federal departments were represented in the POC, Public Safety Canada did not play a salient role in the response.
- CAF: Operation LENTUS 13-01 involved approximately 2,300 regular and reserve personnel from all three branches of CAF. Troops deployed in the seven communities of Calgary, Canmore, Cochrane,

Red Deer, High River, Airdrie, and Medicine Hat. Tasks completed included assisting with the evacuation of civilians,[43] removal of debris, highway remediation work, providing transportation for local officials, water purification, sandbagging, berm construction, culvert installation, and assistance to City of Calgary Emergency Management Agency call centres.
  - 21 June 2013 – 27 June 2013. 6 days.
  - Assets Deployed: 2,200 Army personnel, 100 Air Force personnel, and 15 Navy personnel (a 400 of total personnel were reservists); 6 CH-146 Griffon helicopters; 2 CH-149 Cormorant helicopters; 1 CC-130 Hercules aircraft; 1 CP-140 Aurora aircraft.

## 2015 WILDFIRES | SASKATCHEWAN
Event Details:

- Hazard/Location: Fuel build-up that began in the 1950s along with hot and dry conditions led to vast wildfires across North America in 2015, including an unprecedented stretch of burning forests throughout northern Saskatchewan. Almost half a million hectares had burned by early July, more than thirty times the usual amount of a full regular fire season. More than 100 separate fires raged, affecting a number of northern communities, especially La Ronge and nearby First Nations reserves.
- Impact: The main mechanisms of hazardous impact were the flames themselves, which threatened lives and property, as well as heavy particulate matter in the air caused by wildfire smoke, which can damage some property and have major health consequences. Approximately fifty communities were evacuated with 13,000 people leaving their homes. Response costs exceeded $100 million.[44] The provincial government had to change its initial projection of a surplus budget to a deficit of almost $300 million.
- Duration: While the adverse impacts of the wildfires, especially the lingering health concerns related to particulate matter, stretched from mid-summer into early fall, the most intense period – especially in regard to evacuation response – lasted from early to late July.

Response Details:

- Municipalities: La Ronge and surrounding communities, including First Nation reserves, declared states of emergency throughout July, including reissues of previously lifted declarations. While municipal and band governments helped coordinate civic response efforts, requests for aid were quickly made to the province, especially as the need for evacuations became clear.

- Province: Emergency Management and Fire Safety – Saskatchewan Emergency Measures Organization and the province's Office of the Fire Commissioner are closely aligned – coordinated evacuation responses from within the Ministry of Government Relations while the wildfire management branch within the Ministry of Environment provided updates on air quality and fire hotspots. Like Manitoba, Saskatchewan does not explicitly follow a "whole of government" approach (i.e., peripheral departments are not kept in the loop on response details and provincial coordination occurs within specific ministries and between the premier and the ministers of those ministries, versus at the cabinet table). On 4 July, the Saskatchewan premier spoke to the prime minister regarding the need for CAF support to suppress fires while the province remained in control of evacuation.
- Federal Government: The prime minister responded to the premier's request for CAF, which deployed two days after their conversation. As the fires began to diminish, the prime minister toured the area and spoke to emergency crews in La Ronge. The minister of public safety and emergency preparedness provided updates on the overall response.
- CAF: Operation LENTUS 15-02 involved approximately 850 regular and reserve personnel from the Army and the Air Force. Tasks completed included conducting fire line operations such as patrols, surveillance, digging, and control near cities and critical infrastructure; suppressing hotspots in and near vulnerable communities; and providing logistic support.
   o Duration: 6 July 2015 – 21 July 2015. Total of 15 days.
   o Assets Deployed: 850 personnel; 230 vehicles; 2 CH-146 Griffon helicopters.

## Conclusion

The assessment of interorganizational collaboration between CAF and civilian levels of government during domestic disaster response requires (1) a framework developed for such an assessment; and (2) a qualitative approach that utilizes case studies of disaster events and interviews civilian and military participants who managed those events. With this established, the next step is to apply this framework and methodology in order to determine the effectiveness of large-scale disaster response in Canada.

# 3 The *Presence* of Interorganizational Collaboration

## Introduction

This descriptive part of the overall results paints a picture of the presence of interorganizational collaboration during disaster response before the quality of, and barriers to, such collaboration are assessed in later chapters. The components that constitute the presence of interorganizational collaboration – (1) information sharing, (2) non-manipulative influence, (3) flexibility, (4) support, and (5) collective conflict resolution – are addressed in turn. Each event is compared and contrasted within each component to provide an overall picture of large-scale domestic disaster response in Canada.[1]

Interorganizational collaboration is not only present in contemporary Canadian disaster response but is a primary feature of such response. Indicators of all five components appeared across all events.[2] There are important indicators of the presence of interorganizational collaboration, however, that were not as salient. Hazard type, municipal emergency management capacity, reservist officers, institutional constraints, and specific types of support all played a role in varying the degree of information sharing, non-manipulative influence, flexibility, support, and collective conflict resolution that occurred during disaster response.

## Information Sharing

The indictors for information sharing – distribution of emergency response plans, representatives from multiple organizations in Emergency Operations Centres (EOCs), multi-organizational presence at decision-making tables, the maintenance of communication during response actions, the distribution of new data as events unfold, the

The *Presence* of Interorganizational Collaboration    63

description of organizational strengths and vulnerabilities, and access to a variety of individuals from organizations – were salient features of each disaster response. This section identifies CAF liaison officers as essential to maintaining such robust levels of information sharing and explains how the slight variations that did exist in levels of information sharing across disaster events were due to hazard type, municipal emergency management capacity, and the influence of reservist officers.[3]

*Liaison Officers as Conduits*

By the time each disaster event occurred, Joint Task Force Atlantic (JTFA) and Joint Task Force West (JTFW) had CAF liaison officers (LOs) regularly attend whenever an Atlantic or Prairie province's EOC was "stood up" for multi-stakeholder meetings, or monitor or manage an incident. Such institutionalized participation created a direct link between CAF and the heart of a province's emergency management response system during routine times when hazards were either merely developing into potential disasters or when no significant hazards were present at all. The LO presence in – and contribution to – provincial EOCs was part of the established emergency management process in each province and was not initiated by a formal Request for Assistance (RFA) from provincial to federal governments. Indeed, by 2010, LOs were so entrenched in the day-to-day operational details of provincial emergency management that their commanders had to "remind them that they don't work for the province" (Interview 5, 2017).[4]

The LO concept is well established in CAF doctrine as a conduit in the form of a person that supplies both their own unit and a central headquarters – or, indeed, an operations centre – with relevant information about the other. The goal of an LO in a combat context is to ensure situational awareness is not only increased but comparable across task-defining and task-implementing organizations. CAF commanders saw the importance of such a goal in disaster response and formally applied the LO concept to the domestic context when the regional Joint Task Forces (JTFs) were established with the responsibility to support Canadian civilian authorities within discrete geographic regions. The institutionalized provincial–CAF link as represented by LO participation in provincial EOCs was the result of local networking initiatives by CAF and recipient provinces. Not only did regional JTFs go to great lengths to demonstrate the value of LOs in maintaining situational awareness and enhancing communication flow across organizations, but they also leveraged relationships with former military personnel working in provincial emergency management to make their case (Interview 14, 2017). Becoming established as permanent members within provincial EOCs was no small feat, as even key provincial emergency management stakeholders, from important organizations in the emergency management policy formulation process to provincially contracted providers of emergency services, were not automatically invited to sit in the EOC or attend regular EOC meetings.[5] The following sentiment was common across provincial emergency management leadership: "[For the high level meetings] we did not bother bringing in the Red Cross or Social Services" (Interview 6, 2017). Yet while such provincial players were absent or only part-time attendees of provincial EOCs, CAF LOs – through direct regional JTF–provincial EMO engagement – had developed into permanent members by 2010. The Government of Canada, which holds a constitutional monopoly on armed forces in the country, was not a key player in establishing a routine link between CAF and provincial EOCs.

The routine presence of LOs in provincial EOCs meant that all the indicators for information sharing were salient throughout the military and civilian interviews. LOs were privy to provincial emergency response plan development and, as strategic decisions took place at the EOC, were present at the emergency management decision-making table. The response phase saw a plethora of LOs fuse into the response system, with the main LO in the provincial EOC – or, in the case of Hurricane Igor, the regional EOC in Halifax – sending live updates to LOs embedded within military units deployed across the hazard-impacted area, which allowed an in-person update on the overall civilian response during military "O-groups" as they occurred in each unit.[6]

In some cases, an LO actually drove to platoon commanders overseeing troops on the ground measuring water receding rates or filling sandbags to provide them with province-wide situational awareness. Between personal updates from an LO, military commanders would receive Emails and/or texts from LOs as relevant information arrived at the EOC. Furthermore, the LO in the provincial EOC was instrumental in (1) alerting military commanders of provincial weaknesses that could require military amelioration; (2) indicating to civilian emergency managers that a planned response action was outside of the RFA and hence the military's scope; and (3) linking relevant civilian and military individuals (i.e., an infrastructure and transportation manager with the officer in charge of the deployed engineers).[7] In other words, maintenance of communication during the response actions, distribution of new data as the disaster event unfolded, description of organizational strengths and vulnerabilities, and access to a variety of individuals from another organization were all present during Hurricane Igor, the Manitoba and Alberta floods, and the Saskatchewan wildfires due to the institutionalized CAF–provincial link established during routine and non-routine times through the permanent role of CAF LOs in provincial EOCs.

## Wildfires: The Complicated Hazard

The variation that did exist in levels of information sharing can be explained in part by hazard type. While the suddenly warming weather systems over the Canadian Rockies and Alberta's varying elevation levels meant that the 2013 multi-river flood was fast and violent compared to the slow creep of Manitoba's Assiniboine River flood in 2011, the attendant impacts of both hazards were nevertheless similar: flooded homes and public infrastructure that required sandbagging, road clearing and re-routing, measuring water for receding rates and sewage, evacuating citizens, managing temporary return visits for evacuees, and planning for eventual recovery. Similarly, Hurricane Igor's impacts were largely water based, with the nature of Newfoundland's geography simply requiring more bridge-building to allow transport. The level of technical expertise required of individual responders was therefore general enough to keep the civilian location of the main LO within the provincial EOC; there was no need for the main LO to link with and relay highly specific technical expertise to commanders overseeing troops on the ground. For example, infantry soldiers are trained first and foremost to fight, but in Alberta and Manitoba they leveraged their physical fitness and organizational structures to fill and stack sandbags. Bridge-building in Newfoundland required a higher level of technical expertise, but one easily maintained in-house by the military's engineering trade.

Fighting wildfires, however, required a level of scientific understanding of fires and technical skill that was – aside from military firefighters responsible for fire safety on bases – not readily available within the military. Not only did the soldiers deployed to Saskatchewan require a two-day crash course in firefighting, but the CAF response was largely "assigned" to the province's Wildland Fire Management branch. The allocation of CAF resources to a hazard-specific unit – versus only to an all-hazards and general provincial EMO – meant that the LO had less of an overview of the entire civilian and provincial response and was more focused on the information required to reduce a specific hazard.

While LOs during the floods and hurricane events relayed information relevant to evacuation processes, the psychosocial state of communities, other response organizations such as non-profits and law enforcement, and the sense of managing individuals and municipalities from the provincial EOC, the nature of wildland fire turned the LO into more of an on-the-fly firefighting program accreditor that needed to confirm to JTFW whether soldiers were ready to effectively fight fires.[8] The overall and general levels of information sharing between CAF and provincial officials during the Saskatchewan wildfires therefore lost out in a trade-off for more technical information.

### "Local" LOs

Municipal emergency management capacity also in part explains the slight variation that existed in levels of information sharing. Capacity at the local level determined the scalability of the LO role and hence whether information flowed directly between municipalities and CAF or whether municipalities saw a delay in information flow relevant to CAF support (as the information had to be relayed by the province). Varying levels of municipal emergency management capacity, defined by the degree to which a municipality could perform an emergency management function without external support,[9] meant that (1) a municipality was self-sufficient enough to decline provincial support, which by definition means not requiring CAF support; or (2) a municipality required external support, but was able to run its EOC and manage the resources that are provided within its jurisdiction. Given the scope of disasters that require CAF support, the first of these two scenarios is rare. The City of Brandon, however, developed such an extensive local emergency management network in the decade leading up to 2011 that the Province of Manitoba was not engaged to provide substantial support, despite some of the worst flooding occurring in Brandon. The Brandon EOC was made up of local stakeholders from the city – the non-profit and private sectors – who managed the response with very little CAF contact.[10] Brandon had "managed to have the whole city as an emergency response team," where "everyone had a role to play" (Interview 2, 2017). Brandon's experience was, however, an outlier in the context of this research. Most municipalities could not perform any EM functions without external support, and those that could were large cities nevertheless constrained to managing the resources provided to aid a response (versus producing all the resources themselves).

The City of Calgary had a robust emergency management program and, while not able to manage the 2013 floods without external support, ran response actions that occurred within the city from a state-of-the-art municipal EOC. The technological capacity and established processes of Calgary's EOC were sophisticated enough to house a CAF LO, which allowed a direct link between the municipality and CAF units responding within the municipality. The combination of already decentralized EM in Alberta with the direct Calgary–CAF link manifested a governance phenomenon that is peculiar in the light of Canada's history and constitution, which has inhibited robust federal–municipal links at the forefront of policy implementation: The 2013 floods in Alberta saw the City of Calgary and the ultimate representation of federal power, the armed forces, jointly implement emergency management policy independent of salient provincial guidance. Indeed, one of the provincial ministers responsible for the overall response noted that "we didn't direct them ... they could manage themselves" (Interview 8, 2017). Robust municipal emergency management capacity influenced the scalability of the CAF LO role and thereby the level of information sharing that allowed a municipality to manage resources, including military resources, from within the municipality.

## Reservists and Informal Networks

The final feature that led to variability in otherwise generally high levels of information sharing across the disasters was the influence of reservist officers. With some rare exceptions due to temporary secondments or professional development opportunities, reservist officers were, by their nature, permanent residents of the region where a disaster occurred. Whether they were civilians serving part-time or soldiers with full-time contracts, they were defined by being exempt from the regular force requirement to move wherever CAF needed resources. Long-term residence in an area allowed the building up of networks that is difficult when individuals are periodically moved to new locations. Established networks, in turn, make for easier communication flow during a crisis when the speed of identifying the right people to contact and obtaining the correct contact information is paramount to effective response.[11] The implication of such networks for information sharing during CAF–civilian disaster response was obvious across all responses.

As Hurricane Igor's potential for horrific impact and therefore the need for substantial external support became clear to Newfoundland and Labrador's Fire and Emergency Services personnel,[12] the possibility of requesting CAF support was being vigorously downplayed by the province's premier, Danny Williams, who was in the middle of a long-standing political dispute with Prime Minister Stephen Harper.[13] As Premier Williams had had no qualms about removing all Canadian flags from provincial buildings during a disagreement with a previous federal government, emergency management officials were not certain that disaster response requirements would trump provincial pride. The concern was that the official RFA would be made too late to allow troops to make a substantial impact on protecting lives and property. As statements of a necessary CAF deployment in the provincial EOC (to encourage CAF readiness) would seriously undermine the norm of ministerial – and ultimately premier – authority over decisions that are political in nature (such as requesting military support from another "co-sovereign" under the Canadian constitutional framework), emergency management officials made such statements to peers in their informal network instead. Their peers happened to include reservist CAF LOs and reservist senior CAF officers.

The exact process of how the civilian emergency management network's concerns informally percolated through reservists to JTFA is not clear, but what is clear is that while Premier Williams was still rejecting the notion of help "from Canada," a "movement exercise" was underway from Canadian Forces Base Gagetown in Fredericton, New Brunswick, to Sydney, Nova Scotia, ostensibly as a part of regular testing of CAF's "large troop movement capacity" (Interview 14, 2017). As it turned out, Premier Williams did indeed wait until the worst part of the hurricane had passed before requesting help from the federal government; he submitted an official RFA on 24 September 2010 in the wake of the extensive damage seen across over thirty communities. But given their strategic location just across the shore from Newfoundland in Sydney, CAF deployed in hours and were a salient presence across Newfoundland on the same day.

Alberta emergency management officials likewise relied on their own informal links to JTFW through reservists to increase CAF readiness during the lead-up to the 2013 floods. The delay for an official RFA to be completed within an appropriate emergency management timeline was this time due to wrangling over RFA details within the federal government.[14] As the RFA passed from Public Safety Canada staff to the minister of public safety and then to the minister of national defence and his department's staff, the CAF deployment was – on paper – already twenty-four hours behind when the province would need them on the ground.[15] As in Newfoundland, sending out the "okay to deploy" from the provincial EOC would have circumvented the appropriate political process. Ministers responsible for disaster response anxiously accepted the reality of a delayed CAF deployment to what was fast developing into the worst natural disaster in Canada's history. An emergency management official, however, had touched base with "a contact" at a reservist unit. While a particular minister was "scrambling thinking about the CAF and getting the RFA," they received a phone call from one of their civil service staff who announced that "the military is in Red Deer forward deployed and we need to tell [Premier] Redford" (Interview 8, 2017). After this phone call the minister was "a bit nervous about saying [to the premier] 'Look, this could put you in a slightly difficult position because we haven't followed protocol here. We had to make a decision,' but we briefed the premier at eight o'clock and everybody was fine." The reason emergency managers gave for "everybody being fine" was that CAF readiness, due to the informal links between emergency management civilians and reservist officers, allowed CAF to be deployed when the premier and the prime minister made their first tours of affected areas on 21 June. Government action was not just being promised but was tangibly represented by CAF presence on the ground.[16]

While the impact of reservist officers linked into civilian emergency management networks was clear in the lead-up to Hurricane Igor and the Alberta floods, this link was not tested in the Manitoba Assiniboine River flood nor the Saskatchewan wildfires. Military and civilian research participants from both disasters acknowledged the importance of informal networks, but in both cases the participants noted that a potentially delayed RFA process, which is the problem informal networks address through timely information sharing, was not a concern. Manitoba's successive mass floods over the last couple of decades had institutionalized the RFA process in that province to the point where the sheer amount of provincial and regional federal officials who understood the process meant that a delay on the civilian side was highly unlikely. The relatively frequent flooding in Canada's central province meant that a normal rate of staff turnover was not enough to erase institutional memory of the RFA process in Manitoba. Furthermore, the nature of Manitoban floods – predictable and slow water creep – means that an RFA process in that province is rarely trying to catch up to a hazard the development or impact of which is tougher to predict, such as wildfires, flash floods, or hurricanes.[17] During the Saskatchewan wildfires, in turn, the aforementioned need to train troops in firefighting before they hit the front line was well understood across levels of government and communicated to the media early on. As such, CAF was not seen as the time-sensitive stopgap to the same degree as it was in the other events.

The case for reservist officers as essential to functional informal networks would be stronger if compared to events where mainly regular force officers fulfilled the LO roles and/or to events where local informal links between emergency managers and reservists was not as robust as they were in Alberta and across the Atlantic provinces. However, a detailed assessment of the timeline of events during Hurricane Igor and the 2013 Alberta floods demonstrates the functional nature of such links. As will be further elaborated in the next section, effective disaster response has to find a balance between respecting and circumnavigating legitimate political processes.

### Non-Manipulative Influence and Flexibility[18]

The indicators for non-manipulative influence and flexibility presented similarly across hazards, jurisdictions, and regional JTFs. Two out of the three indicators for non-manipulative influence were salient and one was notably absent across the disaster events studied. In all four cases leaders from CAF, the province, and municipalities had decision-making capacity at decision-making tables as well as the capability to change actions of other organizations in a non-manipulative way. As with information sharing, non-manipulative influence was generally a salient feature during CAF contributions to domestic disaster response. However, the third indicator – a *lack* of institutional constraints to helping disaster response partners – was not salient in any of the events. With some minor exceptions, the RFA – which sets hard parameters around allowable CAF actions – was adhered to from the highest to the lowest ranking CAF member. The RFA is essential to democratic oversight of the military, the efficient management of CAF resources, and – in regard to role clarity and avoiding civilian dependence on CAF support into the recovery phase – even effective emergency management. At the same time, however, the RFA was an institutional constraint to CAF's ability to freely aid disaster response partners in any way its commanders saw fit, as well as civilian – especially municipal – governments' ability to understand the role of CAF during the disaster responses.

### *Breaking the Stereotype: The Military Does Not Take Control*

CAF commanders across all four disasters were amenable to changing their response actions or initial strategies after input from provincial or municipal levels of government. An officer in charge of troops sent in to support the heavily impacted community of High River during the Alberta floods was initially – as per CAF domestic operations doctrine – focused on "life and limb," and wanted to check that any potential evacuation hold-outs present in the town were safe.[19] Upon the mayor prioritizing monitoring of water levels and sandbagging, however, the officer deferred to local knowledge on evacuation rates and evacuee safety and pursued the response actions preferred by the municipal leaders. The officer stressed that "at the end of the day it's their plan. The military does not run the show" (Interview 16, 2017).

During Saskatchewan's wildfires, CAF commanders at joint-planning sessions with the provincial Wildland Fire Management branch planned for the use of Light Armoured Vehicles (LAVs) to move through thick brush. However, even on this issue of tactical strategy (mounted infantry officers have significant experience in assessing suitable ground for LAV transport), CAF deferred to civilian preference for using conventional firefighting transport and techniques to reach hot spots. In Manitoba, sandbagging priorities were identified by municipalities with input from CAF commanders on efficient distribution and use of troops; CAF commanders had no problem removing the City of Brandon from their response plans when that municipality indicated its relative self-sufficiency. In Newfoundland, CAF commanders adjusted their timeline for moving a frigate in order to maintain its visible presence off the shore of a particularly hard-hit community because provincial emergency management leaders had indicated the positive psychological impact the presence of the frigate was having on the community (Interview 1, 2017).[20]

Even in cases where specific CAF response actions – already decided upon by CAF commanders – held obvious public relations benefits to the military, such as the case in Newfoundland when CAF engineers were on hand to quickly fix a bridge holding up a queue of cars, CAF stepped aside to let local contractors perform the action when that was the civilian preference (Interview 1, 2016). Indeed, civilians who worked closely with the military during the response noted that while CAF commanders and soldiers jumped at the opportunity to perform response actions, there was no indication of exploiting the moment for the spotlight; "nobody wanted to play a hero," one minister said of CAF members (Interview 8, 2017).

Unless civilian suggestions to CAF response actions misunderstood a particular CAF capability (i.e., the runway near La Ronde, Saskatchewan was no long enough to allow a Hercules aircraft to land), ran into resource limits CAF commanders on the ground could not control (i.e., helicopters that helped to evacuate numerous Alberta communities could not meet the initial civilian timeline as they were stationed or in use across the country), or requested actions not allowed by the RFA (examples of which are discussed below), CAF commanders were amenable to civilian influence in response actions across the cases studied.

## *The Request for Assistance (RFA) as Institutional Constraint*

The significant civilian ability to make decisions about and influence CAF response actions nevertheless faced the institutional constraint of the formal RFA. Once the RFA passed from a province's solicitor general (or equivalent) to the (federal) minister of public safety, and from there to the minister of national defence for sign off, the parameters for CAF action – their "left and right of arc" in military-speak – were set in stone. Performing actions exclusively within dictated rules of engagement (ROE) is a fundamental pillar of formal CAF doctrine and deeply entrenched in CAF culture, and the RFA is essentially a glorified, large-scale ROE (Interview 16, 2017).[21] This means that civilian – especially municipal – officials at times requested resources and skills CAF had the capacity to provide, but that CAF refused to give as per the RFA. The most common example of a civilian request not granted is related to security support. The officer on the ground in High River during the 2013 floods in Alberta was approached by municipal leaders to provide a security cordon around the evacuated town. While the officer was taken aback by the request given his knowledge of the RFA, municipal leaders thought it only made sense that "the guys in green" run security, especially given strapped law enforcement resources during the disaster (Interview 16, 2017). The officer acknowledged municipal security concerns but stressed that legally he and his platoon were in High River as an Aid to Civil Authority and not as an Aid to Law Enforcement, and that the latter would require enacting different parts of federal legislation and "likely a debate in the House of Commons" (Interview 16, 2017).

The Alberta example is representative of similar events in all the disasters. During natural disaster response on domestic soil, infantry officers – whose actual job descriptions read "to close with and destroy the enemy" – became impromptu custodians of legal propriety in municipalities across the country. Indeed, a civilian emergency management official who worked with CAF at a provincial and municipal level, and in two different provincial jurisdictions, stressed the following:

> A lot of municipalities still have no idea what the heck [the RFA] means. Even a lot of ministers still have no idea that it means that the military will respond [as aids to] civilians during disasters. People have no idea what the military will bring to the table and what they will do and what they won't do because there are parameters. [Municipalities] are looking at increasing security over areas of cities that were evacuated and one of the CAF officers always has to walk over and say "You're not planning on using our members for policing are you?" (Interview 3, 2017)[22]

The RFA constrained the ability of civilian officials to influence CAF response actions beyond security requests. This was most evident when civilian emergency managers and leaders at both the provincial and municipal levels requested CAF actions that were recovery and not response in nature. An appropriately crafted RFA defines in broad terms when an operation's end state is reached, and any actions that can be taken after that state are deemed recovery actions. CAF officers were reluctant to take such actions as both CAF and senior provincial and federal leaders were wary about taking on roles that communities may begin to rely on, but that the CAF could not indefinitely perform. During the Alberta, Manitoba, and Newfoundland events at least one municipal leader requested that troops remove debris and/or pump water from private residences. Such actions were perceived by CAF, and by senior provincial and federal leaders, as appropriately within the realm of individual responsibility and/or municipal capacity implementing their routine public administration apparatus. Similarly, clearing debris from public roads no longer needed for emergency evacuations/services was seen as problematic by some CAF officers as such tasks could be conducted by a municipality's own or contracted services. While a minute level of detail – such as when CAF clearing debris from public roads is permissible – was not always specified in RFAs, the return to the use of routine civilian services as soon as possible and not replicating functioning civilian services throughout the response were clearly indicated.

The distinction between general tasks that were within an RFA's scope (i.e., clearing roads) and replicating existing civilian capabilities, which transcends an RFA's scope (i.e., clearing roads when sufficient and functioning routine civilian services were available), was well understood by CAF officers, but not by most municipal leaders. During the 2011 Assiniboine River flood, the officer in command (OC) of a deployed company gave the order for one his platoons to pull out of a town where water levels had stabilized. A frantic mayor contacted the OC and asked why CAF was getting rid of the town's quick reaction force. "What do you mean 'quick reaction force'?" was the OC's concerned reply. As the need for disaster response actions had dissipated in the community, the mayor had housed the platoon assigned to him at town hall and used them as a general response team to complement or replace thin and tired civilian resources. The mayor's "quick reaction force" were tackling an array of basic municipal service tasks, from supplying extra sandbags to private residences as citizens called in sandbag requests to putting up signage around flood-compromised areas. While CAF fulfilling such tasks made sense to municipal leadership, who understood military aid during disaster as providing general relief wherever needed, the senior civilian officials from the higher levels of government, along with CAF leadership and the RFA's intent, conceived of CAF as a stopgap. These leaders understood emergency management policy as manifested in RFAs to accept temporary municipal pain during the initial recovery phase for the long-term gain of municipal self-reliance. The experiment of the mayor from Manitoba in alternative municipal service delivery was disbanded shortly after the OC learned of it (Interview 15, 2017).

*Strategy versus Tactics: The Flexibility of CAF*

While the ability for municipalities – and mid-level provincial officials – to influence CAF response actions was constrained by the RFA, the discrepancy in understanding the role of CAF during domestic disaster response occurred more between higher and lower members in both military and civilian organizations than between military and civilian organizations as such. Senior CAF, provincial, and federal leaders understood the RFA's constraints, while in the rare circumstances where CAF members did blur the RFA's lines the blurring was done by junior officers. Indeed, the aforementioned mayor's "quick reaction force" was a platoon-level unit and therefore had to be led by a junior officer who, at least for the period of time the "quick reaction force" was active, did not let RFA details get in the way of aiding the community as the mayor saw fit. While varying understandings of RFAs may be a barrier to interorganizational collaboration (which is discussed in chapter 5's section on conceptual difference), the RFA as an institutional constraint to civilian influence over CAF response actions was somewhat tempered when small CAF units overseen by junior officers worked directly with municipal-level officials. Indeed, non-manipulative influence was generally high across all indicators – as opposed to only two indicators – the closer the response actions were to the actual hazard.

The implication the RFA as an institutional constraint held for the presence of interorganizational collaboration was not as straightforward as the implications of high or low levels of information sharing, which indicated a high or low presence, respectively, of collaboration. From an effective disaster response perspective, an institutional constraint that inhibited addressing on-the-ground response needs did not indicate the presence of interorganizational collaboration. From a CAF accountability and democratic perspective, however, legal mechanisms that clearly delineated the scope of military action on domestic soil needed to be upheld. An assessment of CAF's role in all four events demonstrates that the multi-organizational response system had found a balance between the potentially competing requirements of emergency management and democratic accountability. At the senior CAF, provincial, and federal levels, the RFA parameters were set in stone. Any requests for tasks that bubbled up to this level were denied should they even flirt with the RFA parameters or, indeed, the parameters were officially broadened to accommodate multiple requests for a task that could still fall under Aid to Civil Authority – versus Aid to Law Enforcement – with the tweaking of the RFA. This institutional norm ensured democratic accountability won out at the strategic level. Meanwhile, on-the-ground response actions occurring in the midst of fighting a hazard, with junior officers and municipal leaders having to make decisions in the moment, saw effective disaster response win out at the tactical level. Yet the system was self-correcting – leaning towards effective disaster response on the ground never led to junior officers acquiescing to requests that fell under Aid to Law Enforcement. CAF as an organization, then, was remarkably flexible: While all ranks expressed the importance of maintaining democratic accountability within "domestic operations doctrine" *and* achieving effective disaster response, senior officers leaned towards accountability to satisfy their civilian counterparts while junior officers leaned towards effective response to satisfy *their* civilian counterparts.[23]

## Support[24]

CAF's ability to provide and augment resources in Canadian emergency management was substantial across all the events studied. The presence of interorganizational collaboration as assessed by the degree of CAF support was therefore high in domestic disaster response. However, CAF supply was outstripped by the civilian demand in the case of some specific types of resources. The discrepancy between CAF supply and civilian demand for these resources was not affected by hazard type or jurisdiction. This section identifies the many resources CAF was able to provide, the few resources CAF was not able to provide, and CAF resources that were unique to CAF (i.e., no other organization could have provided them).

### *Boots on the Ground*

The greatest CAF contribution to all the disaster responses was human resources. CAF had a number of "sexy capabilities," but ultimately it was the "hundreds of troops with boots on the ground" that provided the greatest relief to affected communities (Interview 7, 2017). Provincial EMOs and other government organizations in Canada had significant capacity in terms of monitoring a hazard's development and treating relatively isolated pockets of adverse impacts (e.g., Manitoba's infrastructure department was hardly stretched thin by having to surveille a particular network of roads more often than usual as water levels rose). "Human power" became a concern when the hazards developed into disasters that impacted so many areas within a jurisdiction that the maximum staffing in that jurisdiction simply could not treat all the issues without quadrupling its usual number of hires (which would often outstrip the number of people available for hire in the first place). Not only were most civilians who were able to work employed in day jobs or attending schools that would not permit long-term disaster response responsibilities, but the pool of civilian individuals physically ready or skilled to perform disaster response actions could not compare to over 1,000 troops with varying trades who arrived with physical fitness and up-to-date medical tests.[25]

Furthermore, even if civilian governments had had the incentive and ability to maximize existing civilian public sector staff through secondments and relevant training, or to establish 1,000-member-strong disaster response teams, they would have had to abide by public sector labour laws and regulations, including the *Canadian Human Rights Act*. Such a non-voluntary civilian disaster response "workforce" would have been entitled to an array of benefits and rights, from overtime pay to accommodation for those who could not perform certain tasks, and, in the highly unlikely event that the workforce was not unionized from the start, would have had the right to unionize, which in turn would have allowed opportunities for the workforce to leverage the disaster event in collective bargaining. CAF members, on the other hand, could not unionize,[26] did not receive overtime pay, and were unique amongst all potential relief workers in that the CAF members could not request accommodation for tasks they could not perform, since CAF's "universality of service" criterion is not subject to the *Canadian Human Rights Act*. The output of CAF workforce was therefore maximized during the disasters' response in a way that would have been legally impossible for a civilian workforce.

No regulations external to CAF dictate how much rest or nourishment soldiers should have; such requirements depend on CAF doctrine, which makes room for commanders' assessment and the mission context. The soldiers that deployed during domestic disaster response slept on average less than four hours every twenty-four hours for the initial seventy-two hours of deployment, which was relatively easygoing compared to the even lower levels of sleep for longer time periods during combat deployments in Afghanistan (Interviews 11 and 16, 2017). Furthermore, soldiers can be ordered into harm's way, up to and including death. There was no comparable civilian organization where managers could direct employees to perform tasks where a jurisdiction's integrity or a community's survival transcended the employee's right to life. Law enforcement and firefighting agencies come close, but – unlike military commanders – their decision-makers cannot explicitly acknowledge a fatality rate and still send their staff in to perform a task. While law enforcement officers, firefighters, and other emergency response organizations accept a risk level, the military accepts a *fatality* level. While no CAF fatalities occurred in the disaster responses studied here, the ethos and legal context that came with the "unlimited liability contract" CAF has with each of its members was present in each case.

The sheer number of troops then – and the scope of their activity given their physical fitness and skill level, and the irrelevance of many labour laws and regulations – meant that CAF could support civilian authorities in immediate and practical ways. Troops dispersed across impacted jurisdictions to clear debris, transport evacuees, check on non-evacuees, transport civilian officials, monitor sewage levels in water, fill and distribute sandbags, build bridges, fight fires, distribute food and water, assess infrastructure integrity, reinforce vulnerable infrastructure, buttress psychosocial support through community engagement, and provide information on all the aforementioned to their CAF chain of command and at municipal meetings they attended. The number of hands-on, practical tasks performed by on-the-ground troops during the first week of each CAF deployment was a key reason why communities were back to a semblance of routine functioning within a month of when the hazards were halted or moved from the area (Interview 17, 2017).[27]

*The Machines: CAF's Hard Assets by Branch*

Physical CAF resources beyond human resources on the ground occurred in the form of branch-specific assets. The Army's main asset across the events studied were LAVs that could transport individuals in water two metres deep and generally navigate rough terrain unmanageable by civilian vehicles (even civilian emergency response vehicles). LAVs allowed platoon commanders to send their sections into the nooks and crannies of flooded regions to facilitate evacuations, check on non-evacuees, distribute food and water, and even transport municipal leaders to isolated residences to engage with community members.[28]

LAVs became as common in the disaster responses as troops on the ground hoisting sandbags or clearing debris. Only in Saskatchewan's wildfires, and then only when moving towards fire hot spots, were LAVs not a dominant part of CAF support to civilian authorities.[29]

The RCAF's main assets were Hercules aircraft for the transport of equipment and materiel; Chinook and/or Sea King helicopters for evacuations; and an assortment of aircraft for monitoring hazard development from the air. Unlike the Army's LAVs, the roles of which were clear across response organizations, and which were easily deployable to the extent that they were needed, CAF's air support faced three separate challenges across the events studied. First, the vast geographic expanse of Canada meant that aircraft were often stationed thousands of kilometres away from disaster sites. Whereas Army troops and LAVs could be on the ground within hours of RFA confirmation, civilian disaster response timelines had to accommodate significant lag time between the RFA being confirmed and the arrival of aircraft.[30] As the Army has a presence throughout Canada (with reservist units and Canadian Rangers in those few regions without significant regular force units, such as British Columbia and the territories), and as the RCN by definition is concentrated on the coasts (albeit with some reservist units scattered across the country), the RCAF was seen by civilian officials and military officers as the CAF branch most disproportionately represented across the country. Manitoba's 2011 Assiniboine River flood was emblematic of the distances RCAF aircraft had to travel to aid the Prairie provinces: Griffon helicopters made the approximately 2,000 km flight from Trenton and Borden, Ontario, and an Aurora patrol aircraft travelled a similar distance from Comox, British Columbia, on the other side of the country. Only Newfoundland could plan for relatively quick helicopter support as Sea Kings were stationed at 12 Wing Shearwater in Nova Scotia.[31]

Second, the capabilities of the various RCAF aircraft were not well understood by civilian officials across all levels of government. Army troops, whether engineers or infanteers, had skills that could easily be linked to concrete tasks by civilian officials, and LAVs mainly allowed the troops and at times civilian officials to move into difficult terrain. In contrast, the contribution of aircraft was felt to be important by all civilian officials, but only those with reservist and/or significant disaster response experience could link a specific craft with a specific task. For example, as the RFAs were being crafted during Hurricane Igor and the Alberta floods, some provincial officials felt that a Hercules was important and wanted to specifically request one as the craft was seen to be potentially useful during evacuations. Hercules aircraft, however, are large, lumbering machines mainly used for transport of heavy equipment, significant numbers of troops, and/or general military materiel. In the rare event that a Hercules aircraft would be used for evacuations, they would require a relatively lengthy runway of hard ground. In other words, a Hercules has to land; it is not going to be plucking individuals from the roofs of houses. Another example was the requests in Manitoba and Saskatchewan for Aurora aircraft to survey the hazard development from above, which – while an accurate assessment of the craft's use – ignored the fact that Transport Canada had similar aircraft better suited to the civilian environment.[32] While CAF domestic operations doctrine and the ideal RFA process eschewed civilian officials linking tasks with specific military capabilities, the reality was that civilian officials did think in terms of capabilities when conceptualizing response plans, and the lack of civilian understanding of the nature of specific aircraft hampered CAF's ability to maximize the effectiveness and efficiency of air support.

Thirdly, the RCAF did not have enough aircraft to fulfil all of its regular responsibilities while responding to domestic disasters. Indeed, the number of military aircraft was the one single area military officers explicitly identified as an inadequate resource during domestic disaster response that could not be compensated for through existing means. Unless Army units were deployed or tasked out, they were conducting training exercises, and domestic operations were leveraged to count as a part of training as all Army trades could practise important aspects of their trades during the domestic disaster response. Even the combat arms trades of the infantry and combat engineers – albeit not the artillery nor the amoured corps – trained their organizational processes and non-combat skill sets, of which there are many (i.e., general labour for the infantry and reinforcing structures for the engineers). The RCAF, however, had so many specific responsibilities in relation to the number of its aircraft that it could not simply replace its operations with disaster response. As the disasters occurred, the RCAF was providing aircraft to NATO deployments to demonstrate Canadian commitment to international security after the Army was pulled out,[33] supporting the ongoing movement of troops and equipment across Canada, maintaining a constant security patrol presence over the second biggest country by landmass on the planet, and – crucially in terms of resource management – operating all search and rescue (SAR) over air in the country. This latter function meant that the RCAF was the only CAF branch a part of the country's routine emergency response system also participating in non-routine disaster response.[34] RCAF aircraft was therefore the one CAF resource that across all cases could not be supplied to match civilian demand.

The RCN's main assets beyond boots on the ground – which were provided largely through inland RCN reservist units – were the Halifax-class frigates that allowed for the ocean-based transport of troops and relief supplies to disaster zones. While frigates were obviously not used or expected in inland disasters, their impact in Newfoundland was substantial. The deployment of troops and relief to isolated coastal communities over the ocean was so effective compared to navigating the hurricane ravaged Newfoundland by land that JTFA allocated three frigates – HMCS *St. John's*, *Montreal*, and *Fredericton* – to the disaster response. While the first round of land soldiers tackled reconstruction tasks to remediate the island's infrastructure, the frigate capability allowed sailors to focus on the relief of individuals. The use of RCN frigates during disaster response meant that immediate CAF support to civilian authorities in Newfoundland was not only physical in nature, but also included a psychosocial element.[35]

*Soft Assets: Tech, Intel, and Funds*

The original collaborative framework developed through the EM, public administration, and civil–military relations literatures identified areas of support beyond physical resources and assets; namely, technological, planning/intelligence, and financial support. The ability of CAF to provide technology support was significantly hampered by the inability of many military and civilian technological tools to "talk to each other," even should they be relatively comparable in sophistication. A prominent example was the common operating picture (COP) tool used by civilian emergency managers and some headquarters-based CAF commanders. A COP is a digital map, akin to Google Maps, of an area that can be "layered" with a variety of relevant details important to strategic decision-makers.[36] For example, a COP in Saskatchewan included developing hot spots and smoke plumes while one in Manitoba showed water coverage and receding water levels. All disaster responses included a COP in at least one EOC. However, civilian COPs (with relevant municipal boundaries) and military COPs (with locations of deployed military units) could rarely "share layers" with each other. They were each announcing one half of the proverbial ball game. Most municipalities, in turn, did not have or could not access COPs at all, despite being physically closest to the details needed to populate them. Indeed, most municipal "COPs" were informal networks and paper maps that provided information and aided decision-making. Ultimately, officers and officials alike had to confirm information through the slower means of calling up individuals on the ground or by physically visiting each other's EOC or headquarters. Similarly, military and civilian radio systems were generally not compatible. Both the military and civilian organizations had comparably sophisticated radio equipment, but the military personnel were in the habit of using crypto (encryption of voice communication over radio waves) and a strictly adhered to system for radio procedure that frustrated the laissez-faire approach to radio communication adopted by (especially non-law enforcement) civilian organizations.[37]

The inability of CAF to provide significant technological support was not seen by any – civilian or military – leaders as problematic to CAF's contribution to domestic disaster response. As face-to-face communication occurred at each leadership level (i.e., junior officers working with mayors and senior officers working with provincial officials), the incompatibility of military versus civilian COPs was at worst a minor irritant, and as tasks once determined were largely conducted by military *or* civilian workers, internal front-line communication was more important than cross-organizational communication, and therefore the gap in radio communication was not seen as adversely impacting the effective performance of front-line response actions (Interview 5, 2017).

The planning support CAF was able to provide across all disaster responses was substantial. Officials involved in different hazards and jurisdictions noted that CAF's constant exercise planning and development during routine, non-deployed periods brought much needed structure and direction within the first seventy-two hours of disaster response (Interviews 3, 4, 5, and 20, 2017). CAF employed a planning process informed by (1) project management methodologies tailored to the military context that the most senior junior officers (i.e., captains in the Army) learn and practise in formal military courses before they can become the most junior senior officers (i.e., majors in the Army); and (2) "battle rhythm" and the orders format, both of which are fine-tuned and rehearsed thousands of times over a commander's career (Interview 16, 2017). Battle rhythm referred to key events that punctuated a set period of time, usually a twenty-four-hour cycle. At a minimum, these events included situation reports (SITREPs) sent from the most junior officers on the ground (which were collated as they moved up the chain of command to inform the situational awareness relied on for planning by senior officers), and a formal O-group to provide updates and define tasks for the upcoming twenty-four hours.

The key feature of battle rhythm that was particularly conducive to planning during the disasters was the steady and predictable timeline of short-term events that were overlaid on a seemingly haphazard and unpredictable developing situation. Regardless of what "the enemy" (i.e., the hazard) did, the battle rhythm chugged along; even if decision-makers could not plan for unpredictable hazard development, they could plan for the next O-group based on the latest SITREPs. The orders format, in turn, was all CAF officers' "holy process" for assessing problems, identifying tasks, and constructing ways of implementing those tasks (Interview 5, 2017). The orders format was the same across all trades and branches; Commanders bundled information according to – and in the order of – the situation, the mission, the execution, the service and support requirements, and the command/signals details. Each bundle had a host of standardized "sub-bundles" for finer detail.

Crucially, the orders format was scalable up and down the chain of command. For example, JTFA command gave orders to its branches, which distilled their particular tasks from the JTFA orders to give orders to companies, which distilled their particular tasks from the company orders to give orders to platoons, which distilled their particular tasks from the company orders to give orders to sections, at which point a single sergeant allocated tasks to individuals that could be traced all the way to regional command, and ultimately to the Canadian Joint Operations Command (CJOC) in Ottawa.[38] The orders format allowed planners to link an individual's action to the highest strategic objectives.[39] This "nature of doing business" in the military was not conducive to all types of civilian planning, but civilian officials who were tackling problems haphazardly prior to CAF deployment generally found stability and structure in CAF's during-response planning process by adopting their own "battle rhythms" and – albeit far less formalized and rigorous – orders formats for task allocation "down the ranks."

The final indicator of support as a component of the presence of interorganizational collaboration is the degree to which CAF contributed financially to disaster response efforts. While CAF was at no point paid directly by provinces/municipalities for any services it provided and never even engaged provinces/municipalities on disaster response budgets, CAF was the main federal instrument used to relieve provincial and municipal burdens during the response. Provincial and municipal governments would have had to spend millions of dollars to hire even a fraction of the CAF resources deployed in the events studied (Interviews 4, 20, 2017). The federal government could theoretically bill the provinces for the CAF response, but the optics of asking a province recovering from a disaster large enough to warrant CAF resources for money, not to mention demanding payment for an event leveraged by federal politicians to demonstrate their commitment to communities, was too problematic for a bill to – at least publicly – materialize.

Officials and officers professed to know of no funds ever being transferred from provinces to the federal government specifically in payment for a CAF response. The CAF contribution to domestic disaster response therefore represented significant financial support across all events, albeit ultimately from the federal government.[40]

### The Greatest Resource: Self-Sufficiency

The single CAF feature that was unanimously lauded as essential to CAF's support to civilian authorities during disaster response was its capability of being entirely self-sufficient. Despite the impact that soldiers on the ground, well-oiled planning processes, and impressive machines have in a disaster zone, all civilian emergency managers stressed CAF's greatest support function was the military's ability to tap none of the already stretched provincial/municipal resources. CAF deployed itself right to the front line, where it clothed and fed and housed its own personnel, and it produced all the resources needed to maintain its operations. Civilian emergency managers acknowledged the importance of what CAF did during the responses, but they stressed what the civilian side *did not have to do* to keep CAF up and running. This orientation by civilian emergency managers illuminated that support was not only perceived as the physical resources provided, but the ability to provide such resources without requiring any support in return. Emergency managers universally identified self-sufficiency, more than any specific resource, as emblematic of CAF's capacity to support civilians during domestic disasters.

### Collective Conflict Resolution

In the rare occasions where conflict occurred between military and civilian members during disaster response, they were explicitly brought to a collective table where the underlying cause of the conflict rather than blame was the focus. The nature of the disaster events, however, combined with the relatively discrete tasks performed by CAF and civilians on the ground, meant the context was not ripe for conflict.[41]

The friction that did occur between military and civilian members occurred at the provincial level in Newfoundland during the initial hours of the deployment and was due to uncertainty around what precisely CAF brought to the table: "There is a natural tendency for some in the provincial authority to think that the military is coming to take over, and the media hypes that as well, but that is not the case. [CAF] is really there as a reassurance piece to back up the provincial authorities" (Interview 1,

Table 3. The Presence of Interorganizational Collaboration

| | NF 2010 | | MB 2011 | | AB 2013 | | SK 2015 | |
|---|---|---|---|---|---|---|---|---|
| | Civilian | Military | Civilian | Military | Civilian | Military | Civilian | Military |
| *The Presence of Interorganizational Collaboration* | | | | | | | | |
| 1. What is the role of CAF in contemporary Canadian disaster response? [descriptive] | | | | | | | | |
| 1.1 Information Sharing | | | | | | | | |
| 1.1.a Distribution of emergency response plans | 2 | 2 | 2 | 2 | 2 | 2 | 2 | 2 |
| 1.1.b Presence in Emergency Operations Centre | 2 | 2 | 2 | 2 | 2 | 2 | 2 | 0 |
| 1.1.c Presence at decision-making table | 2 | 2 | 2 | 2 | 2 | 2 | 2 | 0 |
| 1.1.d Maintenance of communication during response actions | 2 | 2 | 2 | 2 | 2 | 2 | 2 | 2 |
| 1.1.e Distribution of new data as event unfolds | 2 | 2 | 2 | 2 | 2 | 2 | 0 | 2 |
| 1.1.f Description of organizational strengths and vulnerabilities | 2 | 2 | 2 | 2 | 2 | 2 | 2 | 2 |
| 1.1.g Access to a variety of individuals from another organization | 2 | 2 | 2 | 2 | 2 | 2 | 2 | 0 |
| **Result: 104/112 = 93% (HIGH)** | | | | | | | | |
| 1.2 Non-Manipulative Influence | | | | | | | | |
| 1.2.a Decision-making capacity at decision-making tables | 2 | 2 | 2 | 2 | 2 | 2 | 2 | 1 |
| 1.2.b Capability to change actions of others in non-manipulative way | 2 | 2 | 2 | 2 | 2 | 2 | 2 | 1 |
| 1.2.c Lack of institutional constraints to helping disaster response partners | 0 | 0 | 0 | 0 | 0 | 0 | 0 | 0 |
| **Result: 30/48 = 63% (MODERATE)** | | | | | | | | |
| 1.3 Flexibility | | | | | | | | |
| 1.3.a Willingness to reach a decision not originally anticipated by organization | 2 | 2 | 2 | 2 | 2 | 2 | 2 | 1 |
| **Result: 15/16 = 94% (HIGH)** | | | | | | | | |

1.4 Support
1.4.a Any form of resource augmentation (e.g., tech, human, funds, intel) of another's efforts | 1.75 | 1.75 | 1.75 | 1.75
**Result: 7/8 = 88% (HIGH)**

1.5 Collective Conflict Resolution
1.5.a Conflicts explicitly brought to a designated collective table; input of other organizations or a third party sought | 2 | 2 | 2 | 2 | 2 | 2 | 2 | 2
**Result: 16/16 = 100% (HIGH)**
**Total Presence: 86% (HIGH)**

Note: With four participants – two civilian and two military – selected for each disaster response, each indicator could occur a maximum of four times for each event. The maximum of four was reached if the indicator occurred for both civilian participants and both military participants selected for each event. This meant that each indicator could occur a maximum of sixteen times across all of the events. The percentage of each component indicated reflects the number of times all the indicators for a component occurred divided by the maximum amount that an indicator could occur.

2017). CAF and leaders within the Newfoundland EOC addressed initial friction by starting to hold regular joint briefings specifically on the role of CAF in the response where questions could be asked and suggestions provided. By the time the Assiniboine River flooded a year later, the joint briefings had made it into regular CAF domestic operations practice; none of the participants from events subsequent to Hurricane Igor noted any significant conflicts that had to be formally addressed as the response phase took shape. CAF participants stressed that mitigating any potential conflict was successfully done through (1) educating civilians on the RFA – and hence clarifying CAF's role – at joint briefings; (2) avoiding involvement in or even acknowledging inter- and intragovernmental disputes; and (3) explicitly incorporating humility into CAF's deportment and interaction with civilians: "The military knows they don't have all the answers all the time. It's about exchange and collaboration and networking and gleaning information from how others may be doing things better. We reinvent ourselves and try to evolve" (Interview 10, 2017).

Officials and officers stressed that conflict between CAF and civilian organizations, and among almost all organizations on the ground, was extremely rare. The severity of the disasters' impacts focused everyone's goals on "helping Canadians," which led to remarkable cohesion and collegiality across organizations on the front line, and between CAF and civilian officials both on the ground and in headquarters/EOCs (Interviews 2 and 6, 2017). By the time the Newfoundland response matured past its first day, the initial friction had developed into a unified high-level strategy that was "more about team spirit than who gets credit," while on-the-ground municipal leaders were opening up their recreation centres, rinks, and libraries to house soldiers: "At the end of the day we were all a part of the home team" (Interview 10, 2017).

## Conclusion

The presence of interorganizational collaboration, while generally high (see Table 3), was less salient when municipalities had weak EM capacity, reservist CAF LOs were not embedded into provincial EOCs, a hazard (such as a wildfire) required skillsets not part of regular military training, the RFA acted as an institutional constraint on potential CAF response actions, and the civilian demand for RCAF aircraft outstripped the supply. Before potential solutions can be provided to address such issues, however, an assessment of the *quality* of interorganizational collaboration is required. Given the role of CAF during domestic disaster response as described in this chapter, how effective is CAF when it takes on this role?

# 4 The *Quality* of Interorganizational Collaboration

## Introduction

The evaluative assessment of CAF–civilian effectiveness paints a more complex picture than the descriptive assessment of CAF's role in domestic disaster response. While the majority of indicators appeared for each component of the presence of interorganizational collaboration, one component of the quality of such collaboration – integration – saw less than half of its indicators appear.[1] Furthermore, indicators stressed by military officers as essential to the component of coordination – a common vision of, and agreement on, actions that will lead to a response's end state – were largely absent from the disaster responses. At the same time, all the indicators of the other evaluative components – satisfaction and trust – were prominent features across all events, dovetailing with the descriptive components of non-manipulative influence and collective conflict resolution. As all indicators appeared for satisfaction and trust, most indicators appeared for coordination, and less than half of the indicators appeared only in the case of integration (i.e., a majority of total potential indicators within the evaluative assessment did occur), the CAF–civilian relationship during domestic disaster response can be characterized as effective, with the caveat that room for improvement exists.[2]

## Coordination

### *Whose Recovery, Which End State?*

One of the key contributions the civil–military relations academic literature and CAF doctrine made to this book's collaborative framework was the inclusion of a common vision of the "end state" across response organizations – and multi-organizational agreement on the actions required to achieve that end state – as an important indicator of coordination and effective disaster response. The historical significance of deploying a military force, and the weight of all the social, psychological, and economic consequences of doing so, has made defining an end state to military activity a fundamental part of all CAF operations. As the military overview in chapter 1 demonstrated, and as military officers confirmed, the notion of indefinite military action in Canada or by the Government of Canada outside of clearly defined operation parameters is not palatable to liberal democratic norms or CAF doctrine.[3] Furthermore, normative concerns aside, officers echoed CAF doctrine by viewing a common vision of an operation's end state as essential to efficiently managing limited military resources and allowing observable measures for when an operation is either successful or has accomplished all that is feasible given a specific operation's context. Lack of a common vision of the end state between civilian and military leaders during the disaster responses studied here adversely impacted the effectiveness of the civil–military relationship.

The collaborative framework's end state indicator tapped into one of the few areas of substantial difference between military and civilian leaders: the civilian willingness to engage in conceptual ambiguity when discussing the allocation of resources over time. The EM literature, EM best practice, and civilian officials accepted disaster response and recovery as separate phases, but with one flowing into the other. Civilian participants held that the change from response to recovery was not to be characterized by a rigid distinction, and that the change was to be observed rather than planned for in advance. Civilian emergency managers stressed that the nature of disaster is such that pre-establishing hard criteria for when the recovery phase begins is an exercise in false security. "Normal" functioning of the community was acknowledged to be the goal of emergency management, but – since a community after "disasters like these are not 'normal' again for a long time" – using the semblance of "routine" or "regular" processes as a barometer for the beginning of the recovery phase was not seen as useful (Interview 8, 2017). Furthermore, civilian emergency managers disagreed amongst themselves about when the response phases in their disasters ended and when the recovery phases began. Perspectives ranged from identifying the recovery phase as having started immediately once the most acute part of the hazard had passed by, to once all individuals were accounted for, to when physical reconstruction of infrastructure began, to when all individuals had returned to their homes, or to when the federal disaster relief funds arrived months or even years later.

The diversity of civilian views in defining the beginning of the recovery phase caused much consternation among military leaders since the end state of their operation was legally, through the Request for Assistance (RFA), and philosophically, through CAF doctrine, constrained to a clearly defined response phase. As hazards developed into disasters in the cases studied here, civilian leaders reflected EM academic literature in their nuanced – and at times esoteric[4] – debates on what recovery could look like. Meanwhile, CAF was concerned with immediately developing specific and observable measures that would indicate the end state of the disaster response phase, and hence the end of their operation.

Different visions of the end state came into sharp focus almost immediately after CAF deployed. In Alberta, one EM agency official was aiming to begin one of the first joint planning sessions by identifying tasks CAF could potentially accomplish, but the "first question from CAF after they arrived was 'Okay, when are we leaving?'" (Interview 16, 2017). A minister involved in overseeing the provincial response was surprised by the military focus at the start of the response on identifying factors that would indicate the end of the response: "One of the first conversations we had was about that we already had to discuss when the Forces would be leaving. [My reaction was] 'Oh wow'" (Interview 8, 2017). Civilian emergency managers were not oriented towards understanding a disaster response operation as first and foremost a discrete operation amongst many operations. Unlike military personnel, civilians were not "parachuted in" from elsewhere to perform a job and get out. While they understood that the response phase would ultimately end, focusing on what that end would look like before doing anything else was a novel process to them.

Civilian reactions similar to those in Alberta occurred across all the other events except for the Saskatchewan wildfires, and civilian and military leaders provided different – albeit not contradictory – explanations when prompted to explain Saskatchewan as an outlier. An official linked to wildfire management in Saskatchewan pointed to the nature of wildfires and firefighting training as congruent with the military understanding of discrete operations that require pre-determined end states. Wildfires afforded a level of clarity to the hazard development that water-based hazards did not. If flames still existed, the hazard was ongoing and response actions – unless beyond an acceptable risk level to responders – were continued. In comparison, the mere presence of water did not mean a specific response action – à la "keep dousing the fire" – was taken. Flood and receding water rates were assessed, their implications debated, and various response actions developed accordingly. Sandbagging mostly occurred before water arrived and – unlike wildfire, the smallest presence of which could lead to hazard growth – stable, standing water after evacuations was not necessarily a growing threat to individuals or property.

Furthermore, firefighting was unlike the general manual labour needed to fight floods. Firefighters are professional responders with an established set of skills and a body of knowledge that applies when unintended fire occurs and stops applying when the flames stop and/or grow too large to be fought (Interview 17, 2017). Indeed, unlike the officials linked to wildfire management, a Saskatchewan official linked to more general aspects of the response (i.e., evacuations, coordinating among agencies, maintaining health services during the event, etc.) did not see the importance of clearly defining end states when the response phase was only developing (Interview 9, 2017). As established responders to specific, non-routine events that adversely disrupt the norm, firefighters were CAF in microcosm in as far as they "deployed" for a specific task and, as such, needed to know in advance when that task would be defined as complete.

Military officers, however, stressed that the reason a common vision of the end state was a greater part of the Saskatchewan response compared to the other cases was that the military had gone to great lengths to educate its civilian emergency management counterparts on the importance of a defined end state to CAF operations. Civilians in Saskatchewan, according to this perspective, were more in tune to a common vision of the end state not because the response was centred on fighting wildfires, but because the 2015 event occurred after an array of other large-scale events. The importance of clear, agreed upon end states had been pushed by CAF liaison officers (LOs) in provincial Emergency Operations Centres (EOCs) since at least Hurricane Igor in 2010 (Interview 19, 2017). Furthermore, while the focus here is on especially large-scale natural disasters warranting significant CAF support, numerous other RFAs have been responded to since federal EM legislation changed in 2007, and LOs have noticed a steady increase in civilian focus on end states over time regardless of hazard type (Interview 19, 2017). Hazard type may have played a role in the degree to which end states were appreciated and agreed to by civilian and military leaders during wildfires, but the frequency of civilian–military interaction over time and military education on processes important to military deployment played a role as well.

Even as civilian emergency managers across the country (outside of the wildfire management officials) were surprised by CAF prioritizing the solidification of an end state when the response itself had hardly begun,

CAF doubled down on communicating the importance of clear end states. Unlike response actions, CAF commanders did not discuss the preferred nature or relevance of end states; end states were essential to CAF deployments and they were to be clear, observable events that indicated the end of the response phase and hence the end of CAF's operation. One CAF commander stressed that "once the critical need passes away, we don't want to build a dependency" (Interview 1, 2017), and many military participants noted the importance of planning for redeploying CAF troops and assets to their home bases. This latter concern was particularly acute for the Joint Task Force Atlantic (JTFA) commander who had authorized a "movement exercise" to pre-emptively deploy troops near Newfoundland in anticipation that a political dispute between the premier and the prime minister would delay the RFA and hence the speed of deployment. The "movement exercise" had mobilized hundreds of troops, and the commander wanted them back on base conducting conventional training the moment they were not needed.

The emphasis CAF commanders placed on clear end states was not met with a corresponding appreciation for such end states by civilian authorities as the disaster responses unfolded. Despite the fact that the first communication from CAF to civilian authorities on end state importance occurred in all events studied here before RFAs were even in their crafting stages (Interviews 1 and 19, 2017), civilian officials in EOCs and on the ground were still surprised by CAF's insistence on determining an end state by the time forces deployed. This discrepancy in levels of appreciation for end states that would clearly indicate CAF's departure had tangible impacts on the CAF–civilian relationship and the disaster responses. During routine emergency events that happened to occur as the disaster responses were winding down, some civilian agencies had become reliant on and had to use CAF logistics support for what were supposed to be civilian responses (Interview 1, 2017). In some cases, CAF could not withdraw its deployed food and laundry services as civilian groups had begun to use them and had no alternatives ready (Interview 5, 2017). And civilian reliance was not only material in nature but psychological as well. Due to lack of clarity around an end state in Newfoundland, "real challenges arrived" in regard to removing the three frigates floating near Newfoundland shores; the boats had become symbols of the Government of Canada's support and removing them without a clear sense that the response was over involved "sensitivities" that were political enough not to fit comfortably within CAF's mandate (Interview 1, 2017).

The military desire to determine a clear end state was an extension of CAF's insistence on providing support only within the constraints of the RFA. As was discussed in chapter 3, CAF commanders – especially senior officers – were hyper-aware of affirming the norm in Canada of democratic control, implementing Canadian EM policy that ultimately sees civilian authorities – and indeed, communities themselves – responsible for recovery and using CAF resources efficiently. The RFAs allowed CAF to pursue these goals through stating explicitly what the CAF could and could not do while on a domestic operation to support civil authorities, but the RFAs stopped short of detailing end state criteria beyond constraining CAF to the response phase.

If CAF commanders wanted to ensure that CAF's reputation as a professional force subject to civilian authority remained intact, that its adherence to Canadian EM policy was faithful, and that its personnel and assets were responsibly used to be ready for the next operation, then they had to nail down end state details themselves. Civilian emergency managers, however, as with the RFA, struggled to narrow the scope of CAF action to anything less than a general capability and willingness to help wherever and for as long as needed, and encouraged any help they could get to ensure that both their reputations and communities emerged from the disaster intact.

While both civilian and military leaders had incentives to pursue almost all the indicators needed for information sharing, non-manipulative influence, flexibility, support, and collective conflict resolution, on these two indicators of coordination – a common vision of the end state and agreement on the actions required to achieve the end state – the incentives pointed civilians and soldiers in different directions. The consequences of such conflicting incentives were not enough to disrupt the overall CAF–civilian dynamic, but they did lead to confusion over priorities and extended the response phase beyond when CAF resources were needed, both of which hampered the overall disaster response effectiveness.

## *Rejecting Hierarchy: Collegial Networks as Preferred Disaster Response Model*

The discrepancy between civil and military organizations in valuing a common vision of a disaster response's end state did not adversely affect the other indicators of coordination. Civilian and military participants across all disasters characterized interorganizational collaboration as more efficient and effective compared to one powerful organization directing others; civil–military networks were built before and maintained during the responses to allow for strong liaison mechanisms, including embedding staff in each other's organizations where possible;

an appreciation for the impact one organization's activities may have on others was demonstrated on both the civilian and military sides; and consultation occurred prior to either civilian or military organizations taking response actions. The military insistence on, and civilian shying away from, developing concrete measures of an end state was unique in that the other indicators of coordination saw no split along civil–military lines.

When surveyed, officers and officials were prompted for their preference for a command and control approach to disaster response, where one powerful organization tasks out other organizations that are legally and materially subordinate to that organization, versus a collaborative approach where organizations work together as equal partners. Unlike the surveys reviewed in Appendix E, where emergency managers – despite case studies finding little support for the effectiveness of strict hierarchy among response organizations – often support a command and control approach, both officers and officials uniformly advocated for a collaborative approach. There was no difference between civilian and military participants in their support of non-hierarchy. A senior CAF officer who led the post-event lessons learned process for JTFA stressed that while a "common vision of the end state was weak and wasn't clearly defined, there was strong agreement that collaboration is the most effective" (Interview 5, 2017). Another officer essentially – and inadvertently – summed up the academic literature on the subject of a single powerful disaster response coordination agency when he stated that "by necessity, emergency management has to be collaborative … I don't know if you could ever build an organization that effectively tasks others" (Interview 1, 2017).

The fact that the military was the biggest single player on the ground once deployed, with significant and visually impressive physical assets at its disposal, did not lull CAF officers into a sense of control; all military participants stressed that the complex demands of disaster response transcended the capabilities of any one organization. An officer responsible for overseeing a highly lauded mission to aid a particularly hard-hit small town during the Alberta floods stressed that "I don't think even the military has the ability to manage disasters exclusively because of all the various components involved" (Interview 15, 2017). The unfeasibility of one organization in control was echoed on the civilian side as well, where even officials from the provincial emergency management organizations (EMOs), the agencies on paper mandated with coordinating disaster response, noted that the more municipalities and CAF were asked for input – versus directed to take response actions decided at the provincial level – the more successful the response actions (Interviews 3, 7, 9, and 16, 2017).

The emphasis on non-hierarchical interorganizational collaboration by all participants was mirrored in the collegial networking that occurred before and during the disasters – networking that was given the credit by participants for allowing the non-hierarchical interorganizational collaboration to occur in the first place. Networking processes occurred not only through formal liaison mechanisms such as embedding CAF LOs in provincial EOCs, but also through more informal processes where individual CAF officers would touch base with individuals in provincial EMOs during regular, non-disaster times to simply "ensure a link existed" (Interview 16, 2017). One Albertan official remembered getting a call out of the blue from a Joint Task Force West (JTFW) officer to check whether the provincial EMO had any extra information on potential flood development above and beyond what was available on the provincial government's website and through the media. When disaster did strike, the same Albertan official and CAF officer initiated an immediate communication between their organizations as the RFA was being crafted (Interview 6, 2017). The importance the EM academic literature places on having individuals from different response organizations personally know each other before an event happens was therefore reflected in practice and particularly emphasized by CAF officers. For example:

> It's easier to know what to do when you've already made contact with people. Some people may think it's an oxymoron that the military cares about relationships, but while General Hillier was very right in saying we are the Canadian Armed Forces, and our job is to kill people and break stuff, and yes, we can do that too, but you know, militaries don't do or achieve anything if they don't make those connections, those relationships, those working understandings with people they have to work with. (Interview 10, 2017)

Both JTFA and JTFW leveraged their annual domestic operations exercises in the lead-up to disasters to allow individual officers – beyond the LOs – to get to know emergency management leaders at the provincial level (Interviews 5 and 14, 2017). CAF used these exercises as much to identify to civilian emergency managers the military staff that would be there to "assist the provinces when things go bump in the night" as to test the military's domestic operations' capability; the exercises were meant for exercising, but also as a means to "establish relationships" (Interview 5, 2017).

The collegial networking that occurred at the CAF–provincial level was replicated at the CAF–federal and CAF–municipal levels, albeit in less robust and slightly different ways. At the federal level, formal Public Safety Canada liaison links to the Department of National Defence (DND), and hence CAF – links that focused purely on domestic emergency management – have become gradually less robust throughout the twenty-first century.[5] Indeed, individuals with EM responsibility within Public Safety Canada only linked in with DND and CAF as hazards were developing into potential disasters.[6] CAF officers' ability, therefore, to network with individuals on the federal emergency management file during non-disaster times had to occur at those civilian–military tables that did engage during non-disaster times, which were almost always made up of national security focused DND personnel, personnel from one of Public Safety Canada's security-related portfolio agencies,[7] and/or security-focused Transport Canada personnel.[8]

CAF–municipal networking, in turn, saw minor formal support with limited routine contact between officials and the non-Army branches. The "land components established linkages with local mayors and wardens or staff as it's natural for the Army to do it, and only then the Air Force and Navy follows suit" (Interview 5, 2017). While the CAF–federal EM links had to rely on an organizational apparatus put in place for security reasons, but was maintained during non-disaster times, the CAF–municipal link was explicitly EM-focused, yet occurred only during the response phase when hazards were already disasters. The CAF–provincial link, however, managed the best of both worlds, where networking was maintained during non-disaster times and was explicitly EM focused. The networking processes that allowed non-hierarchical interorganizational collaboration to be preferred during disaster response reflected the institutional pressures of the Canadian EM system that concentrates EM capacity at the provincial level.

An appreciation for how one organization's activities may impact other organizations was a salient feature across all the events studied. CAF officers on the ground in small towns were sensitive to the fact that the municipalities they were tasked to support viewed a fire or police response with a handful of vehicles as a significant operation, and that Light Armoured Vehicles (LAVs) or Medium Support Vehicle System (MSVS) trucks carrying 150 troops around a community could be overwhelming from a variety of municipal perspectives, whether short-term city planning, day-to-day operations like bylaw enforcement, or public relations. Platoon commanders (or their equivalent for non-infantry trades) were therefore ordered to establish a "go to" local official immediately upon deployment and to meet with that official every morning before heading out on their platoon's tasks. At those morning meetings, the platoon commander would confirm the tasks the municipality was interested in, explain in general terms how and when their platoon would go about pursuing the tasks, and reinforce the reality that the platoon required no logistical municipal support of any kind and that if anything changed the municipality could directly reach the platoon, often through the platoon commander's personal cell phone number. The daily confirmation of where CAF members would be located and what they would be doing was incorporated into CAF's battle rhythm as CAF leadership was highly cognizant that the unexpected arrival of, for example, forty odd troops pouring out of five LAVs in a usually quiet neighbourhood could cause headaches for town councillors, and in turn disrupt the CAF–municipal relationship. The sensitivity to the impact of significant military activity on day-to-day municipal operations was not affected by hazard or jurisdiction type; the aforementioned process of linking platoon commanders with a local official on a daily basis occurred across the country, from High River, Alberta, to Clarenville, Newfoundland (Interviews 1 and 15, 2017).

The appreciation of one organization's impact on others occurred at levels removed from the front line as well, most notably in provincial EOCs where CAF-involved response actions were jointly decided upon by civil and military personnel. The high levels of the presence of interorganizational collaboration depicted in the previous chapter allowed the type of decision-making processes where military and civilian actions were not determined independent of the impact they may have on each other. In Alberta, platoons ordered to check on non-evacuees were instructed to focus on peoples' safety without affirming their choice to defy evacuation orders (as such affirmation would undermine provincial efforts to minimize harm to emergency personnel safety during the response).[9] In Saskatchewan, military and provincial wildfire-fighting teams were kept separate so that potentially conflicting command structures would not confuse individual firefighters or put military team leaders in the awkward position of "ordering" a civilian firefighter to perform an action. In Manitoba, the military agreed with the provincial assessment that CAF resources in the City of Brandon would destabilize an effectively running municipal response, and hence largely stayed out of the city's response efforts. In Newfoundland, CAF engineers were careful not to begin rebuilding bridges without confirming with the province whether or not the relevant bridges were flagged for CAF rebuilding in civilian plans (in order to avoid scenarios where contractors prepared for and allocated resources to a project only to arrive with CAF already in motion).[10]

Consultation prior to taking action was fundamental to the legality of CAF's disaster response, as well as significant to its organizational culture. The formal RFA, which meant that CAF resources could not officially be deployed without a provincial and civilian request, was revered so much by senior CAF leadership that creative tactics – discussed at length in the previous chapter – were taken to ensure that CAF was ready for a response without formally taking response actions before the RFA was cleared. This legal culture of "request before action" flowed into the day-to-day response culture as "consult before action," whether through joint decision-making sessions at provincial levels or during on-the-ground response actions at municipal levels.

It should be noted that "consultations" were often implicit as the response actions were jointly brainstormed, and therefore neither formal nor unilateral in the way that the term "consultation" suggests. Indeed, the civil–military dynamic, as demonstrated in the high levels of information sharing, non-manipulative influence, and collective conflict resolution discussed in the previous chapter, was such that response actions were confirmed at a strategic level through the decision-making process of a unified team with members holding civil and military expertise, versus separate teams needing to consult each other.[11]

**Integration**

*The Complex Component*

The assessment of integration yielded the most complex result of all the evaluative components. While a common sense of purpose and a unity of effort was present across military and civilian organizations for all disaster responses, significant differences along military and civilian lines occurred for the indicators of integrated activities, interoperability of planning processes, and measures of effectiveness. Discrepancies in situational understanding was also a salient feature across the events, although such discrepancies occurred more between the strategic and front-line (i.e., tactical) levels of disaster response than between organizational type.

Unlike most of the components under the presence of interorganizational collaboration, integration – like every other component under the *quality* of interorganizational collaboration – saw similarities and differences between military and civilian organizations play out similarly across all hazard types.[12] As such, the nature of integration, and of the quality of interorganizational collaboration in general, was determined by the nature of the organizations at play.[13]

*The Coexistence of Solidarity and Separation*

The differences between civilians and soldiers in valuing a defined end state were not reflected in similar differences when officials and officers were asked about their organization's general purpose during the events. All participants defined their organization's purpose as helping communities and Canadians when they are struck by disaster. There was an implicit and perhaps unconscious reluctance on the part of both civilian emergency managers and military leaders to narrow the scope of "their purpose" into phases of emergency management or specific tasks. The purpose of disaster response organizations was repeatedly characterized as achieving a civic goal that transcended the specific constraints and mandates of any single organization. As was discussed in the previous section on end states, and as will be demonstrated below on measures of effectiveness, civilian and military respondents had no qualms about identifying civilian–military differences, but the language of "purpose" sparked characterizations that were stubbornly broad and collaborative in nature. While such a result may have methodological implications for the phrasing of interview questions, it also pointed to the harmonizing impact the disaster context had on individuals from different organizational contexts. The joint nature of the responses along with the intensity of the disaster events themselves produced a level of interorganizational solidarity that was harnessed by words such as "purpose," and could only be neutralized through delving into the specifics of the response actions. Such solidarity demonstrated that a sense of a common purpose, despite differences on how to achieve or measure the outcome of that purpose, bridged military and civilian organizations across all the disasters.

Similarly, "a unity of effort" was universally indicated. Officers and officials pointed to high levels of information sharing, non-manipulative influence, and support – discussed in the previous sections – to demonstrate that a unity of effort existed. However, when their assessments were scrutinized for "integrated activities designed to achieve the common objective," the results were more complicated. Unlike the components for the presence of interorganizational collaboration and the indicators of coordination, "integrated activities" required individuals from different organizations to not only share information, behave in non-manipulative ways, and avoid duplication of tasks, but to actually perform tasks together on the front line in a way that saw the blurring of lines between organizational membership. Such integration was rarely seen during the responses studied here. When a Manitoban mayor utilized an infantry platoon as a "quick reaction force," he did not suggest intermingling members of his municipality's emergency responders with the soldiers. When CAF soldiers had been trained as firefighters in Saskatchewan, they formed their own units and did not individually augment civilian units.[14] CAF engineers in Newfoundland were not fused into contractor teams rebuilding bridges; CAF engineers rebuilt other bridges. In Alberta, sandbagging lines were generally either civilian or military, from filling the bags to placing them. The psychological "unity of effort" that was present across the events did not evolve into physically integrated activities.

The lack of integrated activities on the front line was mirrored in the degree to which civilian and military organizations had interoperable planning processes. As discussed in the previous chapter and the section on coordination, joint decision-making processes that collaboratively detailed "the what" part of response actions were fundamental to each disaster response. Military and civilian organizations jointly decided on what desired effects should occur where. However, the detailed "how" part of CAF response actions were conducted separately in military O-groups that occurred at various levels of command and, unlike provincial EOCs and municipal town halls, did not include any representatives from other organizations. Once the areas in need of CAF help had been prioritized, and the effects required in each area had been established by lieutenant-colonels and their EM counterparts in the provincial EOCs, lower-ranking CAF officers overseeing the implementation of response actions (generally majors or captains in command of a company, or their non-infantry equivalents) began their planning process independent of further civilian insight. This dynamic played out in all the provinces. How to fight back flames, ensure water levels did not rise further, re-establish road access, gain information about non-evacuee safety, monitor town council morale, transport civilian leadership, and so on were all effects expected of CAF, the implementation of which were planned for across all hazards and provinces by groups made up of soldiers.

CAF company-level O-groups manifested classic CAF doctrine. For example (utilizing the infantry's organizational structure), the officer-in-command would "chair" the group, the company's second-in-command officer and senior non-commissioned officers (usually at the warrant rank) would provide updates on the internal organizational aspects within their domains, platoon commanders (the junior officers) would provide updates on their platoons' actions within the last twenty-four hours, an intelligence officer (or, if a commissioned member was not allocated to the company, a non-commissioned intelligence operator) would provide general situational awareness, an LO would provide insight into the company's civilian counterpart, and all of them would be tasked out for the next twenty-four hours after any necessary discussion. O-groups followed a highly ritualized process where specific areas would be addressed at specific times by specific individuals. Individuals speaking out of turn or about an area outside of their domain was extremely rare and rarely tolerated when it did occur (a mistake usually committed only by inexperienced junior officers). Civilians without military experience attending an O-group would not know when to contribute, nor indeed which parts of the information discussed applied to them. Moreover, as so much of the information was passed through in military specific acronyms and phrases, the decisions made may not be clear to civilian attendees (Interview 13, 2017). Indeed, part of the role of an LO once he or she returned to a civilian EOC was to translate military decisions from military-speak into phrases non-military members could understand (Interview 20, 2017).

Planning processes during the disaster responses were collaborative in that decision-making was joint in nature, but the planning that went into accomplishing tasks – and hence the actual front-line activities themselves – was not interoperable. Military leaders directly engaged with their civilian counterparts at provincial levels to make strategic decisions, and lower ranking officers "felt the civilian pulse" by regularly linking up with *their* counterparts at municipal levels. As such, the direction of each disaster response was highly collaborative, but the implementation of the disaster response, as assessed by the degree of interoperable planning processes and integrated front-line activities, was not collaborative.

### Different Measures of Effectiveness

Disaster response integration was further complicated by the different measures of effectiveness employed by CAF versus civilian organizations. Differences occurred within four areas: the valuing of end states; "redeployment" versus staying in a community; the role of politics; and the application of lessons learned. CAF culture and doctrine dictated that military leaders could not even begin to assess effectiveness without a defined end state. Indeed, effectiveness in CAF is a function of whether and how a defined end state was reached (Interview 15, 2017).[15] The lack of a defined end state on the civilian side and, hence, a lack of agreed upon actions to reach the end state led to very different measures of effectiveness between CAF and civilian leaders. In Newfoundland, Manitoba, and Alberta, CAF end states focused on "when the water impact on lives and property was no longer increasing," and, as such, CAF leaders considered established dikes and evacuated communities as indicating success (Interviews 9 and 15, 2017). Civilian leaders, however, did not identify the end of the response phase as a "useful end state" and therefore saw high (albeit non-increasing) water levels around infrastructure and evacuated (albeit safe) communities as an untenable status quo and hence not indicative of anything "successful" (Interviews 2, 8, and 9, 2017). Only in Saskatchewan, where the valuing of and agreement on end states was a part of the response (due to similarities between firefighting and military operations), did both civilian and military leaders view the stalling of hazard development, and hence signs of reaching the end of the response phase, as a success.

A related area that split measures of effectiveness along CAF and civilian lines was the military focus on redeployment. Any military operation needs to end with troops redeployed to their home barracks where they can rest and restart the training process to prepare for the next operation. For military commanders, effectiveness was tied to the ability to redeploy troops. As troops were redeployable once they had accomplished what they were sent to do, redeployability was an indicator for success. While redeployment *can* be indicated by time and not success, as is the case during missions when a soldier's "rotation" is over, success does not occur without redeployment.[16] Civilian emergency managers, however, did not face the reality of having to move their staff or key community players to a "home base" once their response tasks were completed; the disaster zone *was* their home base. This dynamic compounded the CAF–civilian difference in assessing effectiveness as CAF leaders measured success by how fast their soldiers were able to leave an area versus civilian leaders who measured success by how fast their communities could re-enter and continue "life as usual" in the same area. As soldier departure relied on the end of the response phase while regular community functioning occurred nearing the end of the recovery phase, the CAF and civilian timelines used to informed effectiveness were never congruent.

While the valuing of end states and redeployment were relevant measures of effectiveness for CAF and largely absent for civilians, the role of politics was the reverse; politics had little bearing on whether CAF leaders deemed a response successful but was an important variable for civilians. Politics impacted civilian measures of effectiveness through an awareness of political consequences for elected officials when selecting response actions. In all provinces, provincial emergency managers were aware that their minister was ultimately responsible for outcomes and as such included public perceptions as a consideration in their decisions. In Alberta and Newfoundland, where Army forces "leaned forward" before the RFA was cleared in order to be ready for deployment, provincial emergency managers raised concerns on how such prepositioning may look for the minister (Interviews 6 and 7, 2017). CAF, in turn, deployed public affairs officers to ensure the media understood the limits of the military's role and to explain why and how the military took response actions; CAF's media strategy was focused on the military's reputation and not explicitly on an elected official's reputation. Indeed, in the Alberta and Newfoundland prepositioning examples, CAF was more concerned about *not* prepositioning, and thereby potentially presenting as a tardy organization when the RFA was finally cleared, than about the reputations of the ministers of national defence or public safety (Interviews 5 and 10, 2017). The CAF's reputational concerns were oriented towards the organization itself while civilian reputational concerns were oriented towards their respective elected officials; this difference meant that leaders from each organization measured effectiveness through different lenses.[17]

Civilian response leaders at the municipal level manifested a fusion of the provincial and military reputational concerns. This occurred because the response coordination leader – often the mayor or reeve – was also the elected official ultimately accountable to the community for response success or failure. Such a doubling of reputational concerns in measuring success occurred in the context of far fewer resources compared to the provinces or CAF. Municipal leaders outside of large cities did not have large staffs of public servants making decisions with consideration for political careers, nor did they have full-time public affairs personal trained in media relations. The experience of mayors was therefore the disaster experience of the municipal level in microcosm: Despite wielding the smallest amount of resources, they experienced the greatest intensity. Measuring effectiveness, influenced as it was by reputational concerns, became an immensely taxing process for many municipal leaders. Unlike provincial emergency managers or CAF officers, a few small municipalities in Alberta and Manitoba saw their mayors relieve themselves of disaster coordination and be replaced by senior – albeit non-elected – leaders such as fire department chiefs from other, less impacted municipalities (Interviews 14 and 15, 2017). In Saskatchewan, a municipal leader who had served for almost a decade – and had planned to run for election again – cancelled re-election plans after the wildfires, despite being lauded for the municipality's actions during the response (Interview 9, 2017).

The final area wherein measures of effectiveness diverged along military and civilian lines was the role of formal "lessons learned" processes. Unlike valuing end states and redeployment, which are endemic to the nature of CAF, and the role of politics, which is endemic to the civilian context, the use of lessons learned processes is not exclusively linked to either the military or civilian governments and therefore may not indicate intractable differences in how the two organizations measure effectiveness. However, CAF did make far greater use of lessons learned processes than its civilian counterparts.

A disaster response was not deemed successful by military leaders if it had not gone through a formal after action review (AAR), nor if lessons from such AARs were not integrated into future plans for similar operations (Interview 5, 2017). A JTFA officer succinctly articulated the latter sentiment: "Lessons are not lessons if they are not applied" (Interview 10, 2017). Furthermore, CAF AARs explicitly included sections on reviewing the effectiveness of the military–civilian dynamic during the response (Interview 14, 2017). Civilian EMOs, in turn, contributed to or were the beneficiaries of provincial reviews of disaster response. However, EMOs had no mandated requirement to integrate and document specific lessons learned into future plans,[18] and in no case did provincial reviews devote sections specifically to assessing the military–civilian dynamic, despite CAF being the largest single disaster responder in terms of assets and boots on the ground. In other words, civilian emergency managers would deem a disaster response successful independent of officially applying the lessons learned through a formal AAR process, while CAF officers could not indicate success without such a process. This meant that despite agreeing on the facts of the disaster response, from the value of property protected to the number of lives saved, CAF and civilian leaders held different perceptions of the level of success attained by response organizations.

*Strategy versus Tactics 2.0*

The final indicator for integration – a common situational understanding across response organizations – was not a consistent feature of disaster response, although differences in situational understanding did not occur along military and civilian lines.[19] Like the degree to which the formal RFA was understood and abided by, the moment-to-moment understanding of the unfolding disaster situation changed depending on whether individuals were at the "strategic" level or at the "tactical" (i.e., front-line or implementation) level. CAF officers in provincial EOCs mirrored the way senior provincial emergency managers understood the situation in that both sets of leaders agreed upon the status of hazard development (i.e., water level rates, wildfire growth, and hurricane strength), the likely impact of the hazard given its status, the damage already incurred due to the hazard, the general geographic areas where CAF resources could be deployed, and the response capacities of different municipalities.[20] In other words, they were all looking at the same picture. Front-line junior officers and municipal leaders, in turn, also agreed upon the aforementioned areas, but their understanding of the developing situation did not always dovetail with the strategic version.

The discrepancy in situational understanding between strategic and tactical actors played out in three ways, two generally benign and one more problematic. First, front-line individuals observed changes through their own senses as changes occurred, and hence a time gap existed between when a junior officer/mayor observed a changing situation and when a senior officer/provincial official registered the change. In Alberta, Manitoba, and Newfoundland, the tactical actors would note that water levels had stalled while strategic actors were still anticipating and planning for dike-building activities. Such timeline-based differences in situational understanding were easily rectified by the "situation report" (SITREP in military speak) process that regularized information flow from junior to senior officers according to the established battle rhythm, and the more informal but still regular information flow from municipal to provincial EOCs.

Second, as the perspective of tactical actors was constrained to what they were experiencing in their particular geographic context, tactical actors tended to either under- or overestimate the intensity of the overall event. In particularly hard-hit small municipalities in the prairies, mayors and reeves thought the disaster was at its height while in reality the provincial response was planning for the onset of the recovery phase and CAF commanders were planning for redeployment (Interviews 8 and 15, 2017). In Atlantic Canada, CAF engineers repairing and/or constructing temporary bridges in areas with overflowing rivers and roads covered with debris anticipated a prolonged deployment while the improving overall picture meant they were redeployed in a matter of days (Interview 5, 2017). On the other end of the spectrum, one unit of CAF firefighters finding themselves no longer in a "hot spot" due to changing wind conditions anticipated reaching the end of their deployment, while the changing wind conditions actually led to increased fire growth in areas closer to communities, leading to continued deployment (Interview 18, 2017). Such "context biases" by tactical actors generally did not lead to adverse response outcomes as information flow from the strategic level to the front line allowed for quick readjustments of expectations, and soldiers in particular were used to training and operational contexts where they were not entitled to the full picture and needed to fulfil orders regardless of their individual understanding of the situation. Indeed, obvious hazard diminishment in their current area did not impact the urgency with which front-line troops prepared for newly ordered response actions (Interview 20, 2017).

The third and more problematic way discrepancies in situational understanding played out was when strategic players overestimated the accuracy of their understanding due to their "bird's eye view" of the disaster. The voluminous information flow from municipalities and deployed CAF units, along with technological "bells and whistles" such as the digital COPs and hazard modelling software, meant that senior officers and provincial officials felt that their situational understanding was, ultimately, the one to be used to dictate response actions. The problem with such confidence, above and beyond the time lag between front-line events and what is reported in the provincial EOC and military unit headquarters, was that the experience of the individuals fighting the hazard – and hence the truest knowledge of the reality of the disaster response – could not be perfectly reproduced at a strategic level.

In a small town in Alberta, a junior officer had stressed the deteriorating morale and decision-making capacity of municipal leadership, but no replacement actions were taken by the strategic actors as the overall picture was that the floods had stabilized and the general updates from municipalities were positive. When a key member of the deteriorating municipal leadership was finally no longer able to stay on the disaster coordination job, a member from another municipality had to step in, leaving the latter municipality with a gap in leadership (Interview 16, 2017). The provincial government had enough coordination expertise that such a gap at an already stretched municipal level could have been avoided. In Manitoba and Newfoundland, the strategic level pointed to the dissipating hazard and began to "narrow the arcs" on what response actions CAF could undertake, even though in some communities the discontinued actions – such as sandbagging private residences and infrastructure – were still necessary. This led to platoons having to deny community members requesting help for homes and businesses that were facing continued flooding, could be saved with a platoon's labour power, and that could not yet be helped by the municipality because the junior officers had no authority to provide such help (Interviews 9 and 10, 2017). In Saskatchewan, front-line educational briefs on the civilian Incident Command System (ICS) versus the military chain of command system were halted when strategic players noted a large number of briefs had been conducted. Front-line officers and officials, however, stressed that on-the-ground communication suffered as a mutual understanding of the two systems had not yet reached a critical threshold amongst troops and civilian firefighters (Interview 18, 2017).[21]

Discrepancy in strategic versus tactical situational awareness meant that the effectiveness – or at times even the accomplishment – of tasks that were squarely within the response phase were inhibited. Despite the apparent sophistication of the physical, technological, and cognitive maps shared by senior CAF and provincial leadership to make decisions, no map was a perfect representation of the actual territory.

## Satisfaction and Trust

Unlike the first two components of the quality of interorganizational collaboration, the last two were universally and unambiguously present across all the events studied here. Lack of common visions of end states and low levels of integration on a number of indicators had no impact on the degree to which civilian and military response organizations were satisfied with and trusted each other.

The main indicator for satisfaction – approval of partner organizations' actions – was expressed by all. In Alberta, a civilian employer of a reservist LO who took temporary leave from his civilian job to fight the floods was so impressed by the positive impact CAF had on his community and business that he granted the employee extra time off beyond the end of the floods to continue in the LO role and thereby see the event through the recovery phase (Interview 3, 2017).[22] One of the provincial ministers intimately involved in overseeing the management of the Alberta floods stated that "if I was to rank the CAF contribution out of five with five being excellent and one being not so good, they would get a six" (Interview 8, 2017).[23] In Manitoba, CAF officers queried by the JTFW's Disaster and Response Plans unit about whether they would work with their civilian counterparts again all responded with a succinct "Yes," the simple affirmative indicating certainty in the military context (Interview 11, 2017). A senior wildfire management official from Saskatchewan stated that CAF's "overall response was good ... [when they arrive] you know you have guys who can hold their own" (Interview 18, 2017). In Newfoundland, senior CAF officers praised the civilian response: "While the media focus is on the CADPATs disproportionately, we are happy with the province" (Interview 1, 2017).[24]

All the indicators for trust were also salient across the events studied. A broad willingness to have individuals from other organizations influence the use of another organization's resources was evident as provinces could call for desired effects to be achieved on the ground, which impacted the distribution and use of CAF labour and assets. Civilian emergency managers could not dictate the specific implementation process that would achieve the desired effects, but CAF control over such processes was not seen as a lack of trust because (1) the RFA directed provinces to request for effect, not specific capabilities; and (2) the civilian presence at the implementation level was limited to begin with due to the aforementioned lack of integration at that level (i.e., a potential for trust in the context of an integrated process is required for a lack of trust to be indicated). As noted in the earlier discussion on information sharing, the willingness to share data was high for both military and civilian organizations. At no point was information on hazard development, response capacity, or the status of responding units withheld by either civilian or military organizations.

Finally, the willingness to work with each other again was perhaps the most salient of all trust indicators. There was "no pushback" from provincial or municipal leaders in Alberta on engaging CAF again (Interview 3, 2017). While the Manitoba EMO "does not take requesting federal resources lightly," working with CAF again if an event required it would be welcomed by Manitoban officials (Interview 15, 2017). Similarly, while the sense in Saskatchewan's wildfire management community is that the learning curve in fighting fires is such that firefighters from other parts of Canada should be approached first in future wildfires, the province would be happy to work with CAF again should circumstances require it (Interview 18, 2017). The sentiment in Newfoundland in terms of working with CAF again was the most positive as it was expressed without any qualifications. The material impact of soldiers delving into their own food storage to supply civilian communities running low and the psychological impact of seeing a RCN frigate standing watch just offshore were characterized as events invaluable to future disaster responses. As one observer described it, "The roof might be gone, but Canada's still with me here" (Interview 1, 2017).

## Conclusion

Otherwise generally effective civil–military coordination suffered from different visions of – and hence disagreement on the actions that should lead to – an event's end state. CAF's fixation on identifying measurable indicators to identify the end of the response phase never dovetailed with the civilian focus on the end of the recovery phase as the only legitimate end state. The level of military–civilian integration was even lower than coordination as CAF units and their civilian counterparts

Table 4. The Quality of Interorganizational Collaboration

| | NF 2010 | | MB 2011 | | AB 2013 | | SK 2015 | |
|---|---|---|---|---|---|---|---|---|
| | Civilian | Military | Civilian | Military | Civilian | Military | Civilian | Military |
| *The Quality of Interorganizational Collaboration* | | | | | | | | |
| **2. How effective is the civilian–military relationship during disaster response? [evaluative]** | | | | | | | | |
| 2.1 Coordination | | | | | | | | |
| 2.1.a Common vision of the end state | 0 | 0 | 0 | 0 | 0 | 0 | 0 | 0 |
| 2.1.b General agreement on the actions necessary to achieve the end state | 0 | 0 | 0 | 0 | 0 | 0 | 0 | 0 |
| 2.1.c Understanding that collaboration is more effective and efficient | 2 | 2 | 2 | 2 | 2 | 2 | 2 | 2 |
| 2.1.d Strong liaison mechanisms, including embedding staff where feasible | 2 | 2 | 2 | 2 | 2 | 2 | 2 | 2 |
| 2.1.e Understanding the impact one organization's activities may have on other organizations | 2 | 2 | 2 | 2 | 2 | 2 | 2 | 2 |
| 2.1.f Consultation prior to taking action, when feasible | 2 | 2 | 2 | 2 | 2 | 2 | 2 | 2 |
| **Result: 64/96 = 67% (MODERATE)** | | | | | | | | |
| 2.2 Integration | | | | | | | | |
| 2.2.a Common purpose | 2 | 2 | 2 | 2 | 2 | 2 | 2 | 2 |
| 2.2.b Common situational understanding | 1 | 1 | 1 | 1 | 1 | 1 | 1 | 1 |
| 2.2.c Common or interoperable planning process | 0 | 0 | 0 | 0 | 0 | 0 | 0 | 0 |
| 2.2.d Unity of effort with integrated activities designed to achieve the common objective | 0.5 | 0.5 | 0.5 | 0.5 | 0.5 | 0.5 | 0.5 | 0.5 |
| 2.2.e Common measures of effectiveness | 0 | 0 | 0 | 0 | 0 | 0 | 0 | 0 |
| **Result: 28/80 = 35% (LOW)** | | | | | | | | |
| 2.3 Satisfaction | | | | | | | | |
| 2.3.a Expressed approval of another organization's actions | 2 | 2 | 2 | 2 | 2 | 2 | 2 | 2 |
| **Result: 16/16 = 100% (HIGH)** | | | | | | | | |
| 2.4 Trust | | | | | | | | |
| 2.4.a Willingness to work with another organization again | 2 | 2 | 2 | 2 | 2 | 2 | 2 | 2 |
| 2.4.b Willingness to have an individual from another organization dictate actions (e.g., use of resources) for the organization | 2 | 2 | 2 | 2 | 2 | 2 | 2 | 2 |
| 2.4.c Willingness to share data | 2 | 2 | 2 | 2 | 2 | 2 | 2 | 2 |
| **Result: 48/48 = 100% (HIGH)** | | | | | | | | |
| **Total Quality: 76% (MODERATELY HIGH)** | | | | | | | | |

did not share interoperable planning processes, did not integrate their activities at the implementation level, and measured effectiveness through different lenses. However, the universally and unambiguously high levels of satisfaction and trust between civilian and CAF leaders means that the quality of interorganizational collaboration cannot be characterized generally as low (see Table 4). Indeed, as jurisdiction and hazard type did not influence the appearance of indicators, CAF and its civilian counterparts maintained the level of quality that did exist across floods, hurricanes, wildfires, and the variety of differences that exist across provincial contexts. Now that the presence and quality of interorganizational collaboration has been described and evaluated, the analysis can turn to assessing the barriers to such collaboration.

# 5 The *Barriers* to Interorganizational Collaboration

**Introduction**

This chapter builds on the analysis of effectiveness gaps discussed in the previous chapter by explicitly assessing barriers to interorganizational collaboration in order to determine how disaster response can be improved. Each of the components that constitute barriers to interorganizational collaboration – empire-building, manipulation, distrust, benign incapacity, and conceptual difference – are addressed in turn. This more normative assessment of CAF–civilian collaboration demonstrates that *intentional* barriers along military–civilian lines are absent from domestic disaster response. The lack of such intentional barriers aligns with the indicators that appeared in the preceding results chapters, namely under (1) the descriptive components of information sharing, non-manipulative influence, and collective conflict resolution; and (2) the evaluative components of satisfaction and trust.

Intentional barriers, however, were not entirely absent from the responses; an unexpected finding was that a high degree of empire-building was a salient feature of civilian disaster response organizations that existed within the same levels of government.

Most barriers to interorganizational collaboration that did exist along military–civilian lines, however, were due to benign incapacity and conceptual difference. While the indicators for the "intentional" components of this normative assessment identify contexts where organizational success – implicitly or explicitly – trumps ideal disaster response, the indicators for benign capacity and conceptual difference point to material and psychological realities that constrain response actions, even if ideal disaster response is the primary goal for all organizations involved. As with the quality of interorganizational collaboration, the barriers to interorganizational collaboration were largely shaped by the nature of the organizations themselves, and not by jurisdiction or hazard type.

**Empire-Building**

*Organizational Selflessness: Military–Civilian and Intergovernmental Teamwork*

The indicators for empire-building were absent from the CAF–civilian response actions across all disasters. Neither information nor resources beneficial to the disaster response were withheld to enhance the clout of a military or civilian organization, and response actions were tied not to a specific organization's success but to the disaster response itself.[1]

Each province had geographical areas where front-line CAF units were privy to the latest hazard impact and development information, and hence where the CAF chain of command could theoretically have directed the pursuit of response actions independent of civilian input, allowing the increase of CAF versus civilian response accomplishments. However, new information from situation reports sent from front-line units were shared by officers in command (OIC) of companies with their civilian counterparts to allow joint decision-making on response actions. As elaborated in detail in chapter 3, such sharing occurred directly with municipal officials, with provincial officials via a CAF liaison officer (LO) present in the provincial Emergency Operations Centre (EOC), and with provincial politicians via the chain of command (as designated senior officers at the division/JTF level engaged directly with ministers involved in the disaster response). CAF officers across all the events stressed that their role was to – within the constraints of the Request for Assistance (RFA) – provide support wherever civilian officials requested it. The key message conveyed from CAF to civilian leaders when describing CAF capacity can be neatly summed up as follows: "This is what we can bring to the table in support of you. In support of you. Underlined" (Interview 10, 2017). CAF's orientation was towards plugging holes identified by civilian authorities (versus finding opportunities for CAF to pursue response actions), and, as such, the notion of withholding information to support CAF-specific goals was not even on military decision-makers' radar, never mind an explicit option that was contemplated and decided against.[2]

In a similar vein, CAF resources were not withheld from civilian authorities in order to empower CAF's prestige or presence during the disaster responses. As the subject matter experts on effective use of military resources, CAF commanders dictated the specific capabilities and assets required to achieve an effect desired by civilian authorities, but the available "arsenal" from which CAF could develop response actions was constrained only by the RFA and actual limits on resource availability (Interviews 5, 11, 14, and 16, 2017). Specific assets and capabilities were selected purely based on how well they could achieve the requested effects. Unlike combat contexts, where the best equipment may be reserved for CAF members (versus friendly soldiers supported by CAF), or where operational security may dictate keeping the most sophisticated equipment and personnel out of sight, the Aid to Civil Authority domestic context held no incentives for CAF commanders to withhold resources to prioritize CAF-specific goals. Indeed, officers stressed that the nature of natural disaster response on home soil meant that organization-specific goals did not and should not exist; "beating the hazard" is by definition a multi-organizational goal (Interviews 1, 5, 11, 14, 15, and 16, 2017).

Civilian officials across all events studied affirmed CAF's openness in terms of information flow and resource capacity. Municipal leaders noted the availability of platoon commanders on – at times their personal – cell phones; provincial emergency managers stressed that they felt free to confirm information with LOs; and although CAF officers expressed some frustration at requests for specific capabilities versus general effects, such requests from civilians were met with explanations on available capabilities and on CAF's selections (Interviews 3, 6, and 8, 2017).

The lack of empire-building along military–civilian lines was echoed along intergovernmental lines. In Alberta, officials from the provincial EMO stressed the openness of their relationship with the City of Calgary throughout the multi-river flood. Municipal officials were privy to most of the information, including information on resource allocation, the provincial EMO had (Interview 17, 2017). This dynamic was affirmed by a senior Calgary official (Interview 3, 2017) and was essentially replicated by officials from the City of Brandon and the Government of Manitoba when discussing that province's response to the Assiniboine River flood (Interviews 2 and 15, 2017).[3]

The openness in Alberta and Manitoba was noteworthy given the robustness of Calgary and Brandon's EM programs. Provincial EMOs in Saskatchewan and Newfoundland had little incentive to withhold information or resources in order to be the largest civilian disaster response actor as their size and capacity – compared to the hardest hit municipalities during the wildfires and Hurricane Igor (i.e., small towns) – already made the provincial EMOs the primary response actors. The Alberta and Manitoba EMOs, however, arguably had "competitors" in Calgary and Brandon, respectively, but potential for bureaucratic competition did not materialize into any actual intergovernmental empire-building. Emergency managers at the provincial and municipal levels of government saw each other as a part of the same team.[4]

## The Limits of Disaster Response Solidarity: Competitive Intragovernmental Bureaus

The lack of empire-building along military–civilian and intergovernmental lines was contrasted by the indicators for empire-building that appeared whenever civilian officials elaborated upon the role of disaster response organizations within their own level of government. For example, the provincial health emergency management (HEM) branches and the provincial EMOs "built their own empires" to the detriment of each other. In one case, a provincial EMO would only invite HEM representatives to select meetings where health was the primary focus, despite the fact that more generic meetings discussed information useful to the health system response, and despite the fact that CAF officers, municipal officials, and *health representatives from other levels of government* – usually from the Public Health Agency of Canada (PHAC) – were invited to all disaster response meetings (Interview 9, 2017). In an apparent act of retaliation, HEM kept plans for evacuating dialysis patients within its branch until the evacuation was about to take place, which left the EMO officials scrambling as they were responsible for coordinating parts of the transportation process during evacuations (Interview 9, 2019).

In another case, the provincial EMO pointed to a lack of legislation officially mandating the provincial HEM as a valid reason for not incorporating HEM perspectives into provincial response actions (Interview 15, 2017). Non-profit and social service perspectives, however, neither of which were backed in this case by their own emergency management legislation, were incorporated into the EMO's response actions (Interview 15, 2017). Such empire-building among disaster response organizations did not occur only inside single provincial governments, but among EMOs across different provinces. One senior Joint Task Force Atlantic (JTFA) officer noted that EMOs from the other Atlantic provinces were not "first on the list" to provide support during Hurricane Igor, as each provincial EMO had "different styles" and preferred to share information and resources with CAF and their own municipalities than with each other (Interview 1, 2017).[5]

The withholding of information and resources within the provincial level was replicated at the municipal level, albeit at a smaller, more individualized scale. As most – even large – municipalities did not have emergency management–specific health branches (compared to the provincial HEM branches and PHAC's emergency preparedness units), they did not experience the same type of organization-level empire-building. However, information and resources were withheld from within the municipal level by individuals. The "empires" being built in these situations were not for the prestige of a particular organization, but rather for the livelihoods of specific individuals. There were some cases where junior CAF officers on the ground informed their chains of command that reeves and mayors were overemphasizing the hazards in their own neighbourhoods, which led to hazard reduction activities (i.e., debris clearing, sandbagging) occurring first for the homes of municipal leaders (Interview 15, 2017). While such actions conflicted with CAF's military ethics and ethos in regard to general fairness and impartial disaster response, the very same ethics and ethos prohibited CAF from directly engaging civilian governments on their internal dynamics, leaving junior officers to simply stress to their municipal counterparts what CAF understood to be the accurate hazard development picture (Interviews 15 and 16, 2017).

The potential for empire-building within the federal level of government was limited in the disaster zones because two of the empire-building indicators – withholding resources and tying response actions to one organization's mandate – were essentially irrelevant for the civilian federal organizations in the affected provinces given their low response activity. However, the trend for empire-building within the same level of government was affirmed at the federal level as federal organizations did employ the one tool they had in their toolkit – information – to "outperform" their peer organizations. The main federal organizations that had salient EM-related regional branches were PHAC and Public Safety Canada. Both were materially and institutionally limited from significantly influencing response actions themselves, but both had a mandate to provide updated information to the central Government Operations Centre (GOC) in Ottawa.[6] Regional PHAC and Public Safety representatives participated in a variety of EM working groups, networks, and committees (including the provincial EOCs) and could request information relevant to the disaster response, which would be funnelled to their home department's representatives in the GOC. Despite the requested information ending up and announced in the exact same room in Ottawa, regional PHAC and Public Safety organizations maintained and closely guarded their own discrete information flows back to the capital (Interviews 4 and 13, 2017).

## Understanding Low Levels of Empire-Building within CAF

The presence of empire-building observed within each level of government was not replicated within the military, which can be explained differently by emphasizing either civilian or CAF perspectives. First, civilian officials were either unwilling to elaborate upon, or were simply unaware of, empire-building among CAF units. All the aforementioned instances of empire-building by civilian organizations were identified by individuals who were not members of the reportedly empire-building organization, (though in some cases individuals unintentionally unveiled empire-building within the their own organization at the same time[7]), were indirectly confirmed by members of the reported organization,[8] or were described on multiple occasions by individuals with no incentive to deprecate the empire-building organization.[9] Civilians, however, were reluctant to identify areas of empire-building within the military as CAF was the largest, most visible organization that had arrived to help, and had done so in a tangible way with uniformed soldiers physically working on the ground. In the same way that officials did not speak negatively about paramedics, firefighters, or police officers (regardless of any potential empire-building occurring at the high levels of those organizations),[10] the uniformed, front-line aspect of the military's response – accompanied by the aforementioned high levels of satisfaction with, and trust in, the military – oriented civilians away from explicitly reporting negative organizational attributes about CAF. Furthermore, the previously assessed civilian ignorance of internal military processes (including basic modes of communication), along with the non-interoperability of civilian–military units on the front line, inhibited the degree to which civilians were able to discern CAF units withholding information and resources from each other and/or commanders tying response actions to the goals of their specific unit.

Military perspectives, in turn, pointed to CAF ethos and the nature of officer career progression as explanations for the apparent lack of empire-building among CAF units. CAF's code of conduct and ethics, which are promulgated in doctrine and training for the duration of one's military service, explicitly prohibit soldiers from stating anything that could adversely impact the reputation of CAF,[11] and pre-deployment orders and briefs, including media awareness sessions led by CAF's Public Affairs Officers, reaffirmed protecting CAF's reputation as a crucial part of military ethos in all the events studied here (Interview 20, 2017). The officers were therefore indoctrinated against identifying any explicitly negative attribute such as empire-building in their fellow soldiers or their units.[12]

While CAF solidarity – along with civilian wariness about deprecating CAF and civilian ignorance of military processes – possibly shrouded potential CAF empire-building in this research, the incentives of officer career progression also meant that actual levels of empire-building were likely low. Moving through the ranks depends on a variety of factors that range from highly formalized criteria, such as annual personnel evaluation reviews and military course requirements, to highly subjective criteria, such as a superior's assessment of a soldier's initiative, deportment, and leadership skills. Both formal and subjective criteria are similar, however, in that they incentivized officers to respect the chain of command during disaster response; no points were "won" for circumnavigating or even interpreting an order in a way that could not clearly be linked back to a superior's intent.

Officers met career progression criteria through faithfully implementing orders and not through expanding the impact of their particular units. Ordering a unit to stand down to follow an order was more valuable to an officer's reputation and career than racing to the front line and successfully performing an action (Interviews 5 and 14, 2017). Indeed, even when officers received orders that did not maximize the positive impact of their particular unit they defaulted to those orders. In Alberta, no single platoon commander attempted to monopolize evacuations to "tally up lives saved," and all stood down for replacements the moment orders came through (Interview 16, 2017). In Saskatchewan, orders – not the size of the fire to be fought – dictated the location and action of units (Interview 18, 2017). In Manitoba, there was no pushback from, nor lesser praise given to, those platoons that provided logistics support versus those visibly helping on the front line (Interview 15, 2017). In Newfoundland, engineering units had numerous opportunities to maximize the number of bridges they could build, and to thereby "outperform" other trades, but only orders dictated the number of actual bridges built (Interview 5, 2017).

The logic of promotion and the mobility requirement of the military meant that CAF officers would also rarely be in command of one unit long enough to reap many benefits of unit-specific empire-building. Individual officer efforts based on effectively carrying out ordered tasks, be they aggrandizing or of little consequence to the unit's reputation, were far more effective in receiving praise during the military's formal after action reviews than the outcomes of a particular unit (Interviews 14 and 16, 2017). Apart from extraordinary acts of bravery and/or decisions that in hindsight unveil significant strategic errors in the original orders, "successfully performing a unit action" that was not ordered would have been a failure in military terms. The building of unit-level empires was therefore anathema to career success. Individual incentives, perhaps ironically given the prioritization of the group over the individual in military training and units, guarded CAF from the level of empire-building seen within the same levels of civilian government.

## Manipulation and Distrust

None of the indicators for manipulation or distrust were present during domestic disaster response. Indicators that may superficially be allocated to the components of manipulation and distrust were – when assessed in the context of the entire theoretical framework – a consequence of previously discussed components. Not only were organizational manipulation and distrust essentially non-existent – which aligns with the previously discussed high levels of civilian–military information sharing, collective conflict resolution, non-manipulative influence, satisfaction, and trust[13] – but they were non-existent even while levels of empire-building within levels of government were high, and while half the components of integration were low.

Neither military commanders nor civilian officials changed or attempted to change the behaviour of other organizations in a veiled manner in order to achieve their own organization's goals over joint goals. As was elaborated when discussing information sharing, non-manipulative influence, coordination, and empire-building, disaster response goals were not distinguished along military and civilian lines, and joint decision-making by military and civilian leaders on which response actions to pursue was a hallmark of the responses. Officials and officers alike saw themselves as part of a broader purpose that mandated and motivated multiple organizations to aid the collective disaster response. Furthermore, as the broad requests for the effects to be achieved in the disaster zone were legally a civilian responsibility (and accepted in CAF culture and doctrine as such), the military had little incentive to influence civilian organizations in a way that would help CAF dictate the desired effects. Indeed, most of the few instances where CAF did attempt to change civilian behaviour – through *un*veiled education processes – were focused on affirming to civilian officials waiting for CAF to lead the fray that civilians were ultimately in charge and had to make the requests for effects.[14] Civilian organizations, in turn, did withhold information to outperform civilian counterparts within the same level of government. However, as such actions were passive, versus active attempts to adversely influence the behaviour of their counterparts, they were certainly empire-building, but not manipulative.

Indicators for distrust were equally absent from domestic disaster response. Robust informal networks (such as the reservist–EMO links and local officials personally calling platoon commanders working within their municipalities for updates), along with robust formal processes (such as embedding CAF LOs in provincial EOCs and ensuring the relevant JTF allocated a senior officer to work with ministers responsible for overseeing disaster responses), demonstrated that information was not held within civilian or military camps. Neither officials nor officers were caught off guard by assets or actions on the front line because information had not been shared. The examples of civilian organizations within the same level of government withholding information from each other was due to a desire to "outperform" each other and not due to distrust.[15] Indeed, only when the aforementioned opportunities for empire-building arose was information withheld; when no such opportunities existed (i.e., when the EMO in one province was too large to be threatened by that province's HEM, or when another province's HEM had collected information that HEM itself could not act on, or when PHAC or Public Safety Canada had content expertise that was not relevant to *new* disaster updates to be sent to the GOC), information was widely shared even within the same level of government.

No organizations displayed an unwillingness to leave response actions to others. Due to CAF's rigorous implementation of the RFA, along with the lack of incentives that individual officers had to expand their unit's "footprint," the military's default assumption was that a civilian organization would perform a response action unless the joint civil–military decision-making process, informed by civilian requests for effect, had allocated an action to the military (Interviews 1, 4, 15, and 20, 2017). JTFA commanders did not argue for CAF engineers to have a monopoly on bridge-building in Newfoundland; bridges allocated for civilian mending were simply noted as such and passed by (Interview 5, 2017). The JTFW headquarters in Manitoba was happy to leave the City of Brandon to pursue its own response actions (Interview 5, 2017). Military firefighters made no claim to pursue hot spots allocated to civilian units in Saskatchewan (Interview 17, 2017). In Alberta, evacuations were left to only civilians until CAF evacuation support was specifically requested (Interview 16, 2017).

The clarity of CAF domestic operations doctrine on respecting the end of the response phase as an end state to the military contribution also meant that officers were actively encouraging civilian organizations to undertake actions that were recovery-focused. Sandbagging of private residences once water levels had stabilized were characterized by the JTFW in Alberta and Manitoba as the responsibility of property owners and/or municipalities (Interviews 15 and 16, 2017). Once the winds had died down over a number of days and modelling demonstrated the likely shrinking of wildfires in Saskatchewan, CAF leaders there queried when civilian firefighters could hold the line by themselves and – since the immediate threat to communities had been inhibited – whether other civilian firefighters from across the country could be called up (Interview 17, 2017). The JTFA was eager to hand food provision responsibility to Newfoundland and Labrador or individual municipalities as soon as possible to avoid having sailors continuing to share rations – an emergency social services function – from their frigate as aid to communities (Interview 20, 2017).[16] Civilian officials, in turn, supported CAF undertaking an array of response actions, which was demonstrated by the ubiquitous civilian habit of requesting CAF actions beyond both the RFA's parameters and the response phase.[17]

The final indicator for distrust – a preference for working alone – was also absent from the events studied. As Hurricane Igor was developing off the west coast of Africa, as concerned flood forecasters were pointing to high levels of snow build-up in Manitoba and Alberta, and as Saskatchewan's wildfire management community identified dangerously dry wildland patches, CAF LOs were working directly with provincial officials in monitoring hazard development. By the time RFAs were being discussed, the regional JTFs were engaging provincial ministers and, in Alberta and Newfoundland, working – at times informally – with civilian officials to pre-position troops for RFA confirmation. Throughout the actual deployments, CAF officers and their provincial and municipal counterparts worked shoulder to shoulder in sharing information, joint-decision making, collectively resolving conflicts, and coordinating response action allocations. Even when disagreeing on identifying concrete end states, civilian and military leaders were "disagreeing together" by addressing such identifications at joint working groups. The only time civilian and military units worked separately was on the front line due to non-interoperable planning and on-the-ground communication processes. Front-line separation of civilians and soldiers was not, however, "a preference for working alone" and therefore indicative of distrust, but rather a pursuit of the most efficient means of accomplishing the physical tasks of disaster response (Interviews 18 and 20, 2017).[18]

**Benign Incapacity**

Benign incapacity, defined as non-collaboration due to lack of resources, is used in this research to indicate response situations where the desire for interorganizational collaboration was evident but ultimately inhibited by one organization's limited technical, fiscal, institutional, equipment, or labour capacity. Benign incapacity was low across the events studied here as (1) the introduction of the military into the emergency management system by definition infused extraordinary resources into the response landscape, and (2) the fiscal constraints faced by CAF since the Second World War have institutionalized practices and behaviours that maximize outcomes in the context of limited resources. Benign incapacity, however, was not absent from domestic disaster response. Interorganizational collaboration was adversely impacted by limited municipal labour, equipment, and fiscal capacity; limited availability of RCAF assets; and limited institutional support for civilian federal response capacity.

## The Military as Antidote to Benign Incapacity

The scope of disasters that warrant CAF support by definition means that civilian capacity has been overwhelmed. Any amount of CAF resources added to the maximized civilian response was therefore itself an antidote to civilian benign incapacity. The introduction of CAF resources on the ground allowed municipal leaders to allocate less time to organizing their limited front-line response capacity and more time to coordinating with provincial authorities on their municipality's overall response strategy. Such a broader focus allowed mayors and reeves to stop playing metaphorical whack-a-mole with each latest disaster impact in their area and to engage in disaster zone-wide situational awareness and planning (Interviews 6 and 8, 2017). For example, Army platoons checking in on the status of non-evacuees in Alberta removed significant time burdens and stress from small town leaders (Interview 16, 2017),[19] and municipal committees in Manitoba could spend less time organizing local sandbagging efforts once CAF units had deployed, freeing up time for recovery planning efforts (Interview 15, 2017). Likewise, provincial authorities in Saskatchewan were able to allocate less time requesting and coordinating national firefighting augmentation and therefore spend more time working directly with affected municipalities once military firefighters arrived on the scene (Interview 18, 2017). In Newfoundland, the array of small towns and villages heavily hit by Hurricane Igor faced significant inter-municipal communication, coordination, and planning challenges until CAF alleviated the most severe front-line burdens of debris clearing and infrastructure reinforcement (Interview 1, 2017). In all cases, CAF's presence allowed municipalities to move from a purely tactical role to a partly strategic role that more regularly linked them into provincial response processes and thereby allowed greater interorganizational collaboration.

Apart from the alleviation of civilian benign incapacity that CAF by definition brought to large-scale disaster response, the historical context of limited fiscal support of the military in Canada has actually allowed CAF to find creative ways of overcoming resource constraints. As is elaborated in chapter 1's overview of CAF, the Government of Canada has consistently underfunded the country's military compared to Canada's NATO partners, all while participating in international operations at higher rates and levels than proportional to Canada's population and the size of its military. Around the mid-twentieth century, Canada went from being a significant and well-funded military player on the international stage to one rarely funded 2 per cent of Canada's GDP, despite still being mandated with an array of foreign and domestic responsibilities. This reality has meant that CAF members of all ranks have had to be creative to achieve required outputs without the necessary inputs. In a combat context, such creativity has meant that individuals within a specific trade often attain most if not all of the competencies required of that trade. For example, Canadian infanteers have to be proficient on all infantry weapon systems since CAF rarely has the fiscal resources to fund and train weapon-specialized individuals, as the US Armed Forces does. Across CAF's training system – and especially in the reserves – new skills and competencies have been mandated of units not just based on trade but simply on who is available; for example, artillery units have to be able to train for urban search and rescue, and digital image analysis could be handed to units not traditionally digital technology-focused. Despite the military preference for and doctrine emphasizing the importance of working only within "a soldier's arcs," CAF's fiscal reality means that soldiers have to adapt to new responsibilities rather than refine existing responsibilities within established arcs.[20] Such creativity was translated to the non-combat context of natural disaster response.

In Joint Task Forces Atlantic and West, infantry officers – trained first and foremost in orchestrating land combat – relied on short courses and experience to play the LO roles that were so crucial in allowing robust civil–military information sharing, non-manipulative influence, flexibility, collective conflict resolution, coordination, satisfaction, and trust. In Saskatchewan, the CAF norm of quickly "training up" soldiers from a variety of trades to become competent in suddenly required skills allowed CAF to produce functioning firefighters after a rigorous twenty-four-hour course (Interview 18, 2017). Across Manitoba and Alberta, troops without aquatic training or water measurement knowledge took up roles measuring flood levels in wading suits tailored for the purpose (Interviews 15 and 16, 2017). While psychosocial first aid was not a stand-alone CAF service nor incorporated into a specific trade, soldiers – often infanteers – adopted a semi-psychosocial role when engaging with disaster-impacted community members, especially non-evacuees and individuals recently returned to destroyed homes (Interviews 1, 16, and 20, 2017). The military was in general able to leverage the creativity born from the fiscal constraints that define CAF's combat and training reality to mitigate benign incapacity from significantly affecting the military's contribution to domestic disaster response. As one JTFA senior officer put it, "Everyone has a Christmas list [and asks] 'Wouldn't it be nice?' But what we have is what we have. At the end of the day we have who we have, and there likely won't be many more of us" (Interview 10, 2017).

## Municipal and RCAF Resource Constraints

CAF's ability to mitigate benign incapacity – although substantial – was ultimately limited by the resource constraints within the municipal level of government and the Air Force "environment" of the military. Despite the aforementioned dynamic whereby CAF deployment allowed mayors and reeves to pull back from a haphazard tactical role and into a more effective strategic role, the human labour, equipment, and fiscal resources of small municipalities was such that towns and villages remained "consumers" of disaster response resources as opposed to "suppliers." Across the country, local firefighters and law enforcement officials in small municipalities were too few in number to maintain regular functions while supporting disaster response. Even those local companies able to hire extra workers and to expand their hours of operation were limited in acquiring extra equipment necessary to clear debris and build bridges, and most municipalities could not afford to contract out significant extra services beyond what was already covered under operational budgets. While such limited resources do not necessitate low levels of interorganizational collaboration, leaders in charge of "consuming" disaster response organizations by definition had less sway over, and less time to contribute to, multi-organizational groups. Indeed, municipal officials from Calgary and Brandon, both cities significant "suppliers" of disaster response resources within their respective disaster zones, had the time and leverage to be regular players in multi-organizational groups. In other words, "desirable" organizational traits that allow information sharing, non-manipulative influence, flexibility, collective conflict resolution, satisfaction, and trust with other organizations were necessary, but not sufficient, in maximizing collaboration with other response organizations. The ownership of a pool of resources useful to other organizations was crucial as well.

While CAF's significant land assets and creative ability to produce changing support requests through existing resources meant that benign incapacity was largely relevant only to civilian organizations, the RCAF's limited aircraft availability did have minor impacts on interorganizational collaboration. As elaborated in chapter 3's section on support, the RCAF faced three separate challenges during disaster response. First, the vast geographic expanse of Canada meant that aircraft were often stationed thousands of kilometres away from disaster zones. The RCAF was seen by civilians and soldiers alike as the CAF branch with the thinnest dispersion across the country. Second, the capabilities of the various RCAF aircraft were not well understood by civilian officials across all levels of government. While the contribution of aircraft was felt to be important by all civilian officials, only those with military and/or significant disaster response experience could link a specific aircraft with a specific task. Third, the RCAF did not have enough aircraft to fulfil all of its regular responsibilities while responding to domestic disasters. The number of available military aircraft was the one single resource CAF officers explicitly identified as inadequate during domestic disaster response that could not be compensated for through existing means. While the second point above indicates a conceptual difference, the first and third points were purely resource constraints that had an impact on CAF's ability to maximize collaboration with its civilian counterparts.

While CAF commanders actively discouraged civilian officials from linking tasks to specific military capabilities, the reality was that civilian officials did think in terms of capabilities when conceptualizing response plans. Furthermore, the salience of RCAF assets – heightened by the prominent visuals of Hercules aircraft commonly attached to disaster response media stories – and the lack of civilian expertise on the array of capabilities beyond air or even military support available to achieve desired effects, meant that requests for support often included CAF air support.

Civilian requests for specific capabilities that could more effectively and/or efficiently be provided through other means was not unique to air support, but – unlike most land or sea support requests – air support requests put CAF commanders in the arduous position of analysing the availability of an especially scarce resource, and the awkward position of including scarcity in their explanations for their resource allocation decisions. When "Army trucks" were requested in Alberta and Manitoba, CAF commanders stationed there simply confirmed the actual effects desired (non-evacuee check-ins and official transport) and sent Light Armoured Vehicles (LAVs) instead. When "boots on the ground" were requested in isolated Newfoundland towns, CAF commanders also simply confirmed the actual effects desired (water provision and symbolic support), and sent frigates instead. However, when large aircraft were requested in northern Saskatchewan or surveillance aircraft were requested in Alberta, CAF could not simply confirm the effects required (evacuations and hazard monitoring) and provide the "CAF solution" because such solutions – large aircraft that can land on especially short runways and surveillance aircraft – were either already in use or better provided via civilian government organizations. This meant that orienting civilians towards "effects" was not sufficient in developing a response action; CAF commanders had to spend time and effort explaining why CAF was not the ideal or appropriate response actor and suggesting alternative response actors.

Ideal support requests saw civilians request an effect within the formal RFA; less than ideal (but still manageable) requests saw civilians request effects only after CAF officers oriented them away from specific capabilities; and problematic requests saw civilians request capabilities that could not easily be directed towards effects as the effects required could not be achieved by any available CAF capabilities. Only air support requests fell into the last category, where CAF officers oriented civilians towards effects only to still have no CAF capabilities available, which led to CAF redirecting civilians to civilian points of contact for potential solutions. Such redirection was neither efficient nor effective as it used the time of both CAF officers and civilian officials without leading to a concrete military–civilian response action, and it placed CAF officers in the rare position of potentially losing face (and thereby risking the loss of civilian satisfaction and trust). Interorganizational collaboration, therefore, suffered due to CAF's benign incapacity in relation to RCAF resources.

### Federal Institutional Constraints

Benign incapacity also occurred in the non-military federal context as regional Public Safety Canada and PHAC branches were institutionally constrained from influencing response actions. As elaborated upon in chapter 1, EM has – due to Canada's size, decentralized federal system, and provincial authority over policy domains disproportionately affected by disasters – largely become a provincial responsibility.

Furthermore, as municipalities are constitutionally "creatures of the province," the jurisdictions first and most affected by disaster – municipalities – turn first to the provincial level for support. These institutional dynamics had a significant impact on all of the events studied here as both officials and officers agreed that the few federal players that were visible during the responses – Public Safety Canada and PHAC – were glorified messaging systems that mainly relayed information back to Ottawa.

None of the information requested by Public Safety Canada or PHAC could directly be acted upon by either organization in the disaster zones. The implementers of EM policy (i.e., those organizations with the legitimacy to take and/or task out response actions) were the provinces, municipalities, and, within the constraints of the RFA and joint decision-making processes with provincial and municipal authorities, CAF. Despite the significant Government of Canada resources behind them, Public Safety Canada and PHAC were largely response observers with no hard clout to act. A senior JTFA officer summed up the perception that non-federal and military organizations held of civilian federal organizations during the responses:

> CAF [and the municipalities were] responsive to the province, but there wasn't a strong [civilian] federal presence. They did not augment or bring in extra folks ... [they were] thin organizations and mostly there for information to feed info to their chain. The CAF–provincial–municipal team was sufficient. (Interview 1, 2017)

The assumptions within response organizations as well as the response reality on the ground affirmed the low disaster response capacity allocated to the civilian federal level within Canada's governance structure writ large, and the country's emergency management framework in particular. Unlike the differences between provincial jurisdictions, which did not play a role in how disasters were managed, institutions – above and beyond resource capacity and organization type – did matter in regard to the non-military federal role in disaster response.

## Conceptual Difference

Conceptual difference occurred when disaster response organizations held incompatible definitions of the problems that could aggravate and/or solutions that could ameliorate disaster impacts. Such differences between organizations occurred across all disaster responses regardless of jurisdiction type, although the degree of difference was compounded by hazard type in the case of wildfires. This section will first briefly review areas of conceptual difference already indicated under previously assessed components, namely the different CAF versus civilian conceptions of the formal RFA and end states, and the more general conceptual differences that occurred between strategic versus tactical levels of disaster response. The areas of "pure" conceptual difference that were not captured under other components will then be assessed: differences *within* CAF regarding the confirmation process of, and tolerance of front-line troops moving beyond, the RFA; different CAF versus civilian conceptions of solutions to firefighter shortages during wildfires; and different CAF versus civilian conceptions of each other's internal organizational hierarchy when fighting wildfires.

## The RFA and End States 2.0

The two most prominent conceptual differences along military and civilian lines were (1) CAF versus provincial and municipal conceptions of the RFA's constraints, and (2) CAF versus provincial and municipal conceptions of a disaster response's end state. The prominence of these differences was such that they did not only indicate a mere conceptual difference (thereby creating a potential barrier to interorganizational collaboration) but also affected the presence and quality of interorganizational collaboration as well.

Regarding the first point, the lack of "non-manipulative influence" – which is a component of the presence of interorganizational collaboration – was indicated by the RFA being an institutional constraint to CAF freely aiding disaster response partners in any way military leaders saw fit. Civilian governments (especially, but not exclusively, municipal governments) conceived of CAF's role as a broad one that ultimately transcended a written document and could plug any holes during disaster response. CAF leadership, in turn, conceived of CAF's role as *defined* by a written document. To CAF commanders, the formal RFA was essential to democratic oversight of the military, the efficient management of CAF resources, and even – in regard to role clarity and avoiding civilian dependence on CAF support into the recovery phase – effective emergency management. The different conceptions that existed in military versus civilian organizations regarding the legitimate scope of CAF emergency management involvement adversely affected the presence of interorganizational collaboration during disaster response.

Regarding the second point, gaps in "coordination" – which is a component of the quality of interorganizational collaboration – were indicated by disparate visions of a disaster's end state across the events studied. While CAF commanders conceived of well-defined and agreed upon end states as practical necessities and non-abstract conclusions to a disaster's response phase, civilian officials disagreed on what an end state would entail, where an end state would occur along the emergency management cycle, and indeed, whether defining a concrete end state was useful in the first place. The different conceptions that existed in military versus civilian organizations regarding the nature and effectiveness of defined end states adversely impacted the quality of interorganizational collaboration during disaster response.

*Strategy versus Tactics 3.0*

The final conceptual difference that indicated components previously assessed did not occur along military and civilian lines but rather reflected more general divides between the formulation and implementation levels of public policy. "Flexibility," which is a component of the presence of interorganizational collaboration, was indicated by troops and municipal officials at the front line (i.e., the "tactical" level) being more willing to engage in response actions that flirted with the outer edges, or were ambiguously aligned with, the formal RFA. In turn, senior CAF officers, along with high-level provincial officials, defaulted to the "letter of law" when deciding upon response actions. Strategic decision-makers conceived of disaster response as a function of abstract values such as "democratic accountability," while tactical decision-makers conceived of disaster response as fundamentally about the lives, concerns, and property of the people physically in front them.

The strategy versus tactics conceptual difference played out again when the events were assessed for "integration," a component of the quality of interorganizational collaboration. In this case, tactical and strategic individuals understood time differently, as the former conceived of the disaster timeline as moving swiftly, with the hazard's development being observed through their own senses, while the latter conceived of the disaster timeline as steady, with front-line updates arriving at prearranged intervals on a twenty-four-hour cycle or "battle rhythm." Tactical individuals also conceived of the disaster event as accurately represented by the specific disaster impacts they were experiencing in their particular geographic areas, while strategic individuals conceived of the disaster event from a "bird's eye view" and therefore as one that was constitutive of a variety of diverse impacts.

## CAF versus CAF

Two conceptual differences not indicated under previously assessed components occurred within CAF itself. Different perspectives existed across the JTFs with respect to (1) the nature of the RFA confirmation process, and (2) the degree to which front-line actions along or just over the edge of the RFA should be tolerated. Military leaders under JTFW asserted that the RFA confirmation process had been "fine-tuned over the years" so that reality reflected the formal process established in provincial and federal EM policy (Interviews 15 and 16, 2017). According to CAF officers involved in the Alberta and Manitoba events, the provincial EMOs assessed the hazard's impact in relation to provincial resources and, when the latter was found wanting, the need for federal assistance was expressed to the provincial minister responsible for emergency management. From there the minister had the legal right to request assistance via a formal document crafted by the province's solicitor general, which was sent to the federal minister of public safety (MPS). Such a request did not at the outset assume military support; it was up to the MPS to assess the nature of the federal contribution. Once a CAF component was identified as necessary, the MPS engaged the minister of national defence (MND), who cleared the way for the chief of defence staff (CDS) to initiate a CAF response.[21] According to JTFW officers, the CDS then directed JTFW to marshal the force generation capacity of the division in their region and develop a response plan, which was initiated with actual troops on the ground the moment the formal provincial document had been signed by the MPS.

JTFA officers, however, painted a picture of the RFA confirmation process that was neither formal nor unilateral. Indeed, an East Coast LO who was involved in numerous operations of various scales noted that the process to request federal support for provinces during domestic disasters could only be described as "interpretive dance" (Interview 20, 2017). Officers who oversaw the CAF response to Hurricane Igor stressed that the "actual" request for assistance bubbled up the military chain of command even as the provincial EMO engaged its minister on inadequate provincial resources, and the nature of the EMO's interaction with its minister was informed by the CAF LO that sat in the provincial EOC. The minister did not, according to the JTFA perspective, approach the MPS with a generic request for federal assistance, but rather with a specific request for military assistance. The military, in turn, was aware of such specificity, which meant that the CDS – and staff selected to focus on RFAs and domestic operations – was pre-emptively engaging JTFA to prepare CAF response plans while the provincial minister was engaging the MPS (a federal minister with no jurisdiction over the armed forces). By the time the MPS engaged the MND the CDS had received all the necessary information, with the MPS's signing of the provincial RFA understood as a mere formality. While JTFW officers perceived the MPS–MND link as a crucial civilian step in manifesting military involvement in the emergency management domain, JTFA officers perceived the MPS–MND link as little more than a glorified handshake to acknowledge what was already demanded by the disaster reality on the ground.

The conceptual difference between JTFW and JTFA officers could upon first glance be explained by the historical context of their respective disaster events. Hurricane Igor occurred a mere three years after the federal EM legislation and framework was updated in 2007, while responses to the Assiniboine River flood in 2011 and the multi-river Alberta floods in 2013 could learn from the RFA confirmation process followed during Igor. Time and experience, then, and not different JTFs, were potential reasons for why soldiers in Manitoba and Alberta perceived greater allegiance to the RFA process as set out in formal policy. However, the aforementioned JTFA officer with significant domestic operations experience characterized the RFA process as an "interpretive dance" based not only on Igor, but on all the responses they had observed, up to and including the responses to the 2017 Quebec flood and 2017 New Brunswick ice storm (Interview 20, 2017). Furthermore, observations from civilian officials across all the events were aligned with the JTFA conception of the RFA process. Not only did ministers in Alberta engage CAF on pre-positioning military assets before the MPS was formally engaged, but EMOs across the country engaged local and regional CAF units through informal networks – identified and assessed in chapter 3 – before, during, and after the formal RFA confirmation process. Indeed, the reality of the RFA process is far more dynamic and informal than formal policy would suggest. The point in this section is that JTFW officers either believed in, or preferred to express, a formal conception of the RFA process that did not occur in reality and was not shared by their JTFA counterparts.

The other conceptual difference within CAF not captured under previously assessed components also occurred along JTFW–JFTA lines. While all senior CAF officers, along with their senior civilian counterparts, respected the RFA's constraints, senior officers under each JTF held different perspectives on the legitimate amount of toleration to be granted to front-line troops that strayed towards the outer edges of the RFA. For example, when senior officers in Manitoba heard that a platoon was acting as a small town mayor's "quick reaction force" by taking on a variety of local tasks that were the responsibility of the – albeit resource-strapped – municipality, the platoon commander was immediately ordered to cease such actions. Municipal requests that even flirted with the outer edges of the RFA were to be passed up the military chain of command so that senior officers could either assess the validity of the requests or suggest a broadening of the RFA (Interview 15, 2017). JTFW officers had virtually zero tolerance for front-line "interpretation" of areas constrained by the RFA; any ambiguity was to be reported.

In Newfoundland, however, senior officers allowed a degree of toleration for such front-line interpretation while the bounds of the RFA were assessed at a higher level. For example, sailors who shared their own foodstuffs with community members – a function properly under the provincial emergency social services umbrella – were not immediately ordered to stop, even when enough civilian foodstuffs had arrived in the communities (Interview 5, 2017). JTFA leadership sensitive to the RFA did not task out food distribution through a joint decision-making process with senior provincial officials, but when such distribution occurred, front-line decision-makers were allowed to continue until they were explicitly provided with new tasks.

While "mission command" – the military doctrine of allocating decision-making power to the officer immediately in charge of a task – is valued across CAF, JTFA officers defaulted to this value while JTFW officers defaulted to another military value: a hard chain of command. Such differences across the JTFs did not create observable barriers to interorganizational collaboration as civilian officials in each context understood the CAF officers' actions. For example, in Manitoba, provincial officials concurred with the dissolution of the mayor's "quick reaction force" as re-establishing local capacity was the goal of provincial EM policy, and in Newfoundland provincial officials understood the CAF sharing of foodstuffs as more a psychological boost to the community than a replacement of civilian services.

### *Wildfires: The Complicated Hazard 2.0*

Conceptual differences not captured under previously assessed components that occurred along CAF–civilian lines did, however, raise barriers to interorganizational collaboration. Both of these "pure" conceptual differences – "pure" because they were not primarily indicated under the presence and/or quality of interorganizational collaboration – occurred during wildfires. The first difference was anchored in varying civilian versus military conceptions of the other's internal organizational make-up. Civilian firefighters utilized the industry standard Incident Command System (ICS), which identifies a single incident commander and allocates operational, logistical, financial, and other discrete areas of responsibility to specific individuals. The military, in turn, was internally organized by the chain of command, where commissioned officers develop and are responsible for strategy that is implemented by non-commissioned members; at both the commissioned and non-commissioned levels, higher ranks dictate tasks to lower ranks. While on the surface a relatively strict hierarchy was present in both ICS and the chain of command (versus the consensus-building decision-making processes that are more common in non-military and non-emergency management organizations with horizontal internal structures), a key difference existed between the civilian ICS and the military chain of command: In the former, leadership positions were determined by subject matter expertise and/or context, while in the latter, leadership positions were a function of rank.

The ICS did not require the most senior official or the official with the most experience to be the incident commander, while CAF's chain of command automatically allocated the decision-making power to the highest rank available, and in the event of the same rank, experience level held sway. During wildfires, this discrepancy meant that front-line junior officers seeking to confirm response actions with their civilian counterparts approached the most senior provincial official or firefighter in their vicinity, an individual who often was not the incident commander responsible for managing the junior officer's civilian sister units (Interview 18, 2017). Throughout the disaster zone, soldiers struggled with engaging and responding to individual authority that was temporarily appointed versus individual authority that was entrenched through rank, a dynamic that led to unnecessarily delayed and confused communication on the front line.

A counterintuitive mechanism helped explain why the ICS versus chain of command conceptual difference created a barrier to interorganizational collaboration in Saskatchewan. While all provincial EMOs moved into an ICS structure during the response phase, the civilian front lines in Alberta, Manitoba, and Newfoundland were organized loosely and horizontally through consensus-based local non-profits, highly collegial and familiar municipal staff, small companies without rigorous internal hierarchies, and an array of community members not formally organized at all. In such cases, the difference between the military chain of command and the civilian front line was so stark that front-line junior officers allowed their subordinates to engage directly with their civilian front-line counterparts and engaged only the mayor and/or designated local officials when confirming response actions (Interview 16, 2017). In Saskatchewan, however, the civilian front line was almost entirely made up of professional firefighters formally organized. Junior officers recognized what they thought were pseudo-military structures and therefore treated them as such; subordinates could only communicate through the chain of command, and junior officers sought out individuals comparable to themselves in rank to engage in response action confirmation. Senior officials linked to the provincial wildfire management branch and senior firefighters were therefore often approached by junior officers to engage on details about which the officials and firefighters had no situational awareness and on which they had no capacity to act (Interview 18, 2017).

The other conceptual difference that played out in Saskatchewan saw different military versus civilian perspectives on the ideal solution to a shortage of firefighters. While the presence and quality of CAF–civilian collaboration was generally high during the 2015 wildfires, CAF officers and senior provincial officials disagreed on whether training and employing soldiers as firefighters was the most effective approach to the province running out of human resources. CAF officers conceived of the wildfires as any other hazard that had developed into a disaster and saw the Saskatchewan RFA as no different from those initiated across other jurisdictions. Provincial officials, however, conceived of firefighting as a specialized skill with soldiers padding firefighter numbers as an unsatisfactory solution. Ironically, this conceptual difference led to similar CAF and civilian perspectives on lead-up training; CAF leaders were adamant that their soldiers be as proficient in fighting fires as they were in the more mundane skills of debris clearing and sandbagging, and wildfire management officials were adamant that non-firefighters were given significant support to effectively work next to professional firefighting units. Soldiers therefore went through a firefighter program that was jointly developed by wildfire management officials and CAF, and – despite being condensed – yielded results largely satisfactory to civilian firefighters (Interviews 18 and 20, 2017).

"Satisfaction," however, was conceived of differently along CAF–civilian lines. CAF officers initially expected soldiers to be fully functional on the front line, as they were in other hazards, but civilian officials constrained CAF's presence to the "mid-tier" of wildfire fighting, which meant that soldiers were not fighting the largest fires along with the civilian units closest to the hottest zones. Such different conceptions led to some confusion during joint decision-making processes as CAF officers were not used to incorporating geographical limits that did not pertain to civilian units into their planning. The dynamics of

Table 5. The Barriers to Interorganizational Collaboration

| | NF 2010 | | MB 2011 | | AB 2013 | | SK 2015 | |
|---|---|---|---|---|---|---|---|---|
| | Civilian | Military | Civilian | Military | Civilian | Military | Civilian | Military |
| *The Barriers to Interorganizational Collaboration* | | | | | | | | |
| **3. In what ways might the military contribution to Canadian disaster response be improved? [normative]** | | | | | | | | |
| 3.1 Empire-Building | | | | | | | | |
| 3.1.a Withholding information that could benefit joint goals to enhance the power of one organization | 2 | 0 | 2 | 0 | 2 | 0 | 2 | 0 |
| 3.1.b Withholding resources that could benefit joint goals to enhance the power of one organization | 2 | 0 | 2 | 0 | 2 | 0 | 2 | 0 |
| 3.1.c Conceptualizing response actions as tied to the goals of one organization | 2 | 0 | 2 | 0 | 2 | 0 | 2 | 0 |
| **Result: 24/48 = 50% (MODERATE)** | | | | | | | | |
| 3.2 Manipulation | | | | | | | | |
| 3.2.a Changing or attempting to change the behaviour of others in a veiled manner in order achieve an organization's goals over joint goals | 0 | 0 | 0 | 0 | 0 | 0 | 0 | 0 |
| **Result: 0/16 = 0% (LOW)** | | | | | | | | |
| 3.3 Distrust | | | | | | | | |
| 3.3.a Unwillingness to share information | 0 | 0 | 0 | 0 | 0 | 0 | 0 | 0 |
| 3.3.b Unwillingness to leave response actions to others | 0 | 0 | 0 | 0 | 0 | 0 | 0 | 0 |
| 3.3.c Preference for working alone | 0 | 0 | 0 | 0 | 0 | 0 | 0 | 0 |
| **Result: 0/48 = 0% (LOW)** | | | | | | | | |
| 3.4 Benign Incapacity | | | | | | | | |
| 3.4.a Non-collaboration due to a lack of resources (e.g., technical expertise, equipment, funds) | 1 | 1 | 1 | 1 | 1 | 1 | 1 | 1 |

(*Continued*)

|  | NF 2010 | | MB 2011 | | AB 2013 | | SK 2015 | |
| --- | --- | --- | --- | --- | --- | --- | --- | --- |
|  | Civilian | Military | Civilian | Military | Civilian | Military | Civilian | Military |
| **Result: 8/16 = 50% (MODERATE)** | | | | | | | | |
| 3.5 Conceptual Difference | | | | | | | | |
| 3.5.a Incompatible definitions of the problem and/or the solution (i.e., different characterization of the scope of disaster) | 2 | 2 | 2 | 2 | 2 | 2 | 2 | 2 |
| 3.5.b Different ideas of who/what is to blame | 0 | 0 | 0 | 0 | 0 | 0 | 0 | 0 |
| 3.5.c Different views on who can take a response action | 2 | 2 | 2 | 2 | 2 | 2 | 2 | 2 |
| 3.5.d Different appreciations for the best type of organizational model to use in disaster response | 0 | 0 | 0 | 0 | 0 | 0 | 2 | 2 |
| 3.5.e Varying knowledge levels/interpretations of emergency management best practices | 1 | 1 | 1 | 1 | 1 | 1 | 1 | 1 |
| **Result: 44/80 = 55% (MODERATE)** | | | | | | | | |
| **Total Barriers: 31% (LOW)** | | | | | | | | |

such planning did not lend itself to the fluidity seen in other responses, to the point where provincial officials ideally preferred a "surplus of professional firefighters" with whom to make plans. Such a preference, however, did come with a caveat that elevated CAF above extra civilian firefighters: Saskatchewan was the province that received the bulk of the military support during a fire season that affected most of Western Canada as other provinces were allocated foreign firefighters to aid their responses, and officials in Saskatchewan agreed that – based on their knowledge of the responses in other provinces – they would rather fight fires with Canadian soldiers than incorporate firefighters from an array of different countries (Interview 18, 2017).

## Conclusion

Not only were the intentional barriers of empire-building, manipulation, and distrust non-existent when CAF and civilian disaster response organizations worked together, but the military was largely an antidote to the benign incapacity civilian levels of government would have experienced without a CAF contribution. CAF officers' creativity in achieving desired civilian effects meant that all branches except the RCAF were able to deploy resources and assets in ways that significantly reduced the overall disaster response timeline. However, different CAF versus civilian conceptions existed on the legitimacy of RFA constraints, which reduced the presence of CAF–civilian collaboration, as well as on the nature of end states, which reduced the quality of CAF–civilian collaboration. Barriers to CAF–civilian collaboration were also created in Saskatchewan due to different military versus civilian conceptions of each other's internal organizational structures and the ideal solution to firefighter shortages.

The most serious barriers to interorganizational collaboration did not occur along CAF–civilian or intergovernmental lines. Rather, the intentional component of empire-building occurred across all disaster responses for civilian organizations within the same level of government (see Table 5). The implications that such empire-building has for policy-makers and practitioners, together with the other findings of the three results chapters, are assessed in chapter 6.

# 6 Results, Implications, and Recommendations

## Introduction

The concepts of the presence and quality of, and the barriers to, interorganizational collaboration have been treated separately up to this point in order to analyse the components of each in detail as they manifested themselves across disaster responses. All three concepts can now be assessed together to discuss general themes that emerge from an examination of domestic disaster response and to build recommendations on how such responses may be improved in the future.

The crucial role of informal networks; the negative hazard-specific impact of wildfires; the largely positive effects of weak federal disaster response capacity; front-line integration as unnecessary collaboration; intragovernmental competition for resources; conceptual differences surrounding the formal Request for Assistance (RFA) and "end states"; and military–civilian "sharing" of the decision-making role define the nature of disaster response in Canada. It follows that Canadian disaster response can be improved by fostering and expanding emergency management (EM) networks, maintaining decentralized EM in Canada, and enhancing RCAF, municipal, and all-hazards capacity across the country.

## A Collaborative System

An assessment of the results obtained under all three of the concepts developed in this book's theoretical framework demonstrates that the Canadian governmental disaster response system can be characterized as largely collaborative. Over two thirds of all potential indicators for the components under the presence and quality of interorganizational collaboration occurred during disaster response, while less than one

Table 6. Summary of Collated Results

| Research Question | Concept |
| --- | --- |
| 1. What is the role of CAF in contemporary Canadian disaster response? [descriptive] | The *Presence* of Interorganizational Collaboration (86% = High)*<br>**Components:**<br>- Information sharing (93% = High)†<br>- Non-Manipulative influence (63% = Moderate)<br>- Flexibility (94% = High)<br>- Support (88% = High)<br>- Collective conflict resolution (100% = High) |
| 2. How effective is the civilian–military relationship during disaster response? [evaluative] | The *Quality* of Interorganizational Collaboration (76% = High)<br>**Components:**<br>- Coordination (67% = Moderate)<br>- Integration (35% = Low)<br>- Satisfaction (100% = High)<br>- Trust (100% = High) |
| 3. In what ways might the military contribution to Canadian disaster response be improved? [normative] | The *Barriers* to Interorganizational Collaboration (31% = Low)<br>**Components:**<br>- Empire-building (50% = Moderate)<br>- Manipulation (0% = Low)<br>- Distrust (0% = Low)<br>- Benign incapacity (50% = Moderate)<br>- Conceptual difference (55% = Moderate) |

Notes:
* The percentages next to each concept reflect the percentage of total potential indicators that occurred for each concept. For example, across all the responses studied, and considering the total number of research participants, the indicators for all of the components under the presence of interorganizational collaboration could occur a maximum of 200 times; as they occurred 172 times, the percentage allocated to the "presence" concept is 86 per cent (172/200).
† The percentages next to each component reflect the percentage of total potential indicators that occurred for each component. For example, across all the events studied, and considering the total number of research participants, the indicators for information sharing could occur a maximum of 112 times; as they occurred 104 times, the percentage allocated to information sharing is 93 per cent (104/112).

third of the potential indicators for the components under the barriers to interorganizational collaboration were observed (see Table 6).[1]

Just under 90 per cent of all the potential indicators for the *presence* of interorganizational collaboration occurred across all disaster responses, with over 90 per cent of the potential indicators occurring for the components of information sharing, flexibility, and collective conflict resolution. Non-manipulative influence suffered due to the RFA acting as an institutional constraint to helping disaster response partners, but was still a "moderate" presence as 63 per cent of the potential indicators of non-manipulative influence did occur. Support came in just under 90 per cent as RCAF assets were limited across all disasters. In general, the indicators of four out of the five components for the presence of interorganizational collaboration were found just under or over 90 per cent of the time. The presence of interorganizational collaboration was a salient and fundamental part of domestic disaster response.

The *quality* of interorganizational collaboration was not as uniformly high as the presence of such collaboration, but 76 per cent of all potential indicators for quality did occur. Coordination came in at a moderate 67 per cent of potential indicators due to the lack of a common vision of – and agreement on the actions necessary to achieve – a disaster response's end state. Integration further dragged down the overall percentage of potential quality indicators as interoperable planning processes and common measures of effectiveness were not present across military–civilian lines, and there was a discrepancy in situational understanding between "strategic" and "tactical" levels of disaster response in both military and civilian spheres. A common purpose, however, was indicated across civilian and military organizations for all events. Similarly, all the indicators for satisfaction and trust occurred across civilian and military organizations for all disasters. Indeed, if interorganizational perceptions (satisfaction and trust) were separated from interorganizational processes (coordination and integration) under the "quality" concept, then all of the potential indicators for perception occurred across the disaster responses. The quality of interorganizational collaboration was generally high, albeit with room for improvement.

The indicators for the components of the *barriers* to interorganizational collaboration dovetailed with the indicators for the components of the presence and quality of such collaboration; only 31 per cent of potential barrier indicators occurred across the disaster responses. Half of the potential indicators for empire building occurred as intragovernmental – but not *inter*governmental, civilian-military, or intra-military – empire-building occurred across the disaster responses. Manipulation and distrust, however, were essentially non-existent. Indeed, the barriers to civilian–military collaboration were the non-intentional barriers of benign incapacity (with half of potential indicators occurring) and conceptual difference (with just over half of the potential indicators occurring). The level of benign incapacity that existed was due to limited municipal EM capacity and RCAF assets, while conceptual difference occurred due to different understandings of the RFA, the value of end states, and – in the case of the Saskatchewan wildfires – what distinguished an Incident Command System from the military chain of command.

An assessment of the indicators that occurred for the components of all the concepts demonstrate that the dynamics "internal" to the disaster response system (i.e., all the interorganizational perceptions and processes that were not disaster response actions in and of themselves) were highly collaborative in nature. Information sharing, non-manipulative influence, flexibility, collective conflict resolution, satisfaction, and trust were characteristic of all the disaster responses. When such internal dynamics did suffer, they did so mainly due to conceptual differences on the nature of the RFA and the value of end states. The already high level of internal interorganizational collaboration can be improved even more through addressing such differences.

The dynamics "external" to the disaster response system – support, coordination, and integration (i.e., those interorganizational processes directly a part or supportive of disaster response actions) – were more complex in nature. CAF support to achieve disaster response actions was generally – although not universally – high, while federal and municipal support to achieve disaster response actions was generally low.[2] The components for civilian–military coordination were indicated at moderate levels while civilian–military integration was low.[3]

Lack of support was due to benign incapacity while the lack of coordination and integration were due to conceptual difference. Limited RCAF assets were identified in those few instances of low CAF support, which implicated benign incapacity in terms of the constrained military ability to provide air support. As municipalities were by definition "consumers" (rather than "suppliers") of disaster response actions, their limited resources meant that they could not be equal partners in coordinating or providing disaster response along with CAF and the provinces. The civilian federal organizations, in turn, experienced similar benign incapacity as that of the municipalities despite the significant resources of the federal government because they were institutionally constrained by the Canadian EM system in particular, and the structure of Canadian governance in general.

Levels of coordination, in turn, were moderate only due to civilian–military conceptual differences on the value of end states (the other components of coordination were salient across all events). Similarly, the lack of "common measures of effectiveness" under integration was due to civilian–military conceptual differences on the value of end states. The non-interoperability of some planning processes and frontline actions under integration, however, was not inhibited by any of the components under the barriers to interorganizational collaboration, and was due rather to the functional requirements of disaster response. Indeed, as will be addressed in the following section, non-integration in some disaster response domains did not lead to – and may actually be a requirement of – effective disaster response. The external dynamics of interorganizational collaboration that do warrant improvement can be addressed through mitigating conceptual differences to achieve higher levels of coordination and integration, and through enhancing resources (i.e., diminishing benign incapacity) for support.

Interorganizational collaboration was both present and of generally high quality during the disaster responses, which means that CAF was overall effectively integrated into domestic EM during the response phase. However, civilian–military conceptual differences and, to a lesser degree, benign incapacity, meant that such effectiveness could not be maximized.

## General Themes

### The Importance of Informal Networks

Relationships between individuals established prior to disaster events and outside of formal organizational contexts were instrumental to the presence and quality of, and the lack of barriers to, interorganizational collaboration. The formalized role of CAF liaison officers (LOs) in provincial Emergency Operation Centres (EOCs) was necessary but ultimately not sufficient on its own for facilitating the high levels of information sharing that did occur. Such formal roles were informed by informal links LOs had built up with their civilian counterparts either through making connections with former military members in the EM community or getting to know civilian emergency managers outside the EOC during non-disaster times. Indeed, reservist LOs, who had greater opportunities to foster relationships with civilian emergency managers given that reservists reside locally and/or have one foot in a civilian career, allowed for especially robust communication flow during the lead-up to disasters.

The informal links fostered by LOs with their civilian counterparts were merely the tip of a reservist–civilian network iceberg that saw a vast amount of information flow between the provincial EM community and the local CAF units during hazard development. Such networking meant that military units responsible for a disaster zone did not have to wait for their formal chain of command to gain fresh knowledge on whether CAF support might be required. In both Alberta and Newfoundland, delays in the RFA process could have meant a delayed CAF response (resulting in more disaster damage), yet the relevant commanding officers were made aware through informal networks that the likelihood of deployment was high despite RFA delays, which in turn led to prompt CAF deployments the moment the RFAs were confirmed.

Informal networks were not only relevant to high levels of information sharing (thereby improving the presence of interorganizational collaboration), but also allowed for the collegial and horizontal interorganizational processes that improved levels of coordination (thereby enhancing the quality of interorganizational collaboration). Non-hierarchical interorganizational collaboration was a fundamental feature of all the disaster responses and was credited by both soldiers and civilians as essential to maintaining strong liaison mechanisms, an appreciation of one organization's impact on other organizations, and consultation processes prior to action (i.e., all the indicators that did occur for coordination relied to some extent on non-hierarchical interorganizational collaboration). Such non-hierarchical collaboration during the disaster response phases only occurred because collegial relationships between CAF officers and emergency management officials had been fostered through informal networks prior to the disasters.

Informal civilian–military networks built up during non-response phases also meant that CAF officers and EM officials did not associate their response goals with either military or civilian organizations, but rather as goals of the "disaster response team," which encapsulated both CAF and emergency management organizations (EMOs). Military and civilian leaders alike viewed their goals as a part of a broader common purpose, and neither information nor resources beneficial to the disaster response were withheld to enhance the clout of a military versus a civilian organization. Established informal networks meant that an "us versus them" mentality, severely problematic for effective disaster response, was not a feature of the civilian–military response.

The conclusion that informal networks are important affirms by now well-established findings in EM and public administration literatures. As discussed at length in chapter 2, informal networks characterized by collegial collaboration at an individual level and spontaneous (as opposed to formalized) organization can ameliorate a host of misallocation and miscommunication problems that might otherwise occur during a disaster response (Zurcher 1968; Kendra, Wachtendorf, and Quarantelli 2003; Wachtendorf and Kendra 2005; Voorhees 2008; Groenendaal and Helsloot 2016; Boersma et al. 2014; Jensen and Thompson 2016; Rimstad et al. 2014; Jensen and Waugh 2014). Similarly, public administration scholars stress the importance of informal networks that transcend hierarchical, organizational, or even jurisdictional lines when confronted by complex and unpredictable phenomena such as disasters (Agranoff and Mcguire 2003; O'Leary and Bingham 2009;

Simpson 2011). This book confirms that such a general finding holds true for information sharing and coordination processes when civilian and military organizations work together in the Canadian context during domestic disasters.

Emergency management practitioners, in turn, can leverage this finding that informal networks are crucial to effective disaster response by proactively establishing working relationships with individuals throughout their EM communities. By the time a hazard has developed into a disaster and the response phase begins, it is too late to effectively identify the best individuals most likely to have decision-making capacity, the most up-to-date information, the ability to provide last-minute resources, and so on. Such knowledge is latent and ready in whatever informal network has already been developed.

Emergency management best practice should dictate not only establishing formal interorganizational links such as liaison mechanisms and embedding individuals in different organizations' EOCs, but also that individuals expected to play roles during disaster responses should be reaching out to each other informally and regularly during non-disaster times. Indeed, as was aptly demonstrated in Alberta, senior EM officials were able to get an immediate status on CAF readiness as the floods were developing because they called a Joint Task Force West (JTFW) officer who had reached out "just to touch base" a few weeks earlier. Formal plans, processes, and communication algorithms were important during the disasters, but informal connections not formally documented were crucial.

## The Hazard-Specific Effects of Wildfires

Specific hazard characteristics did not affect the presence or quality of, or barriers to, interorganizational collaboration, except in the case of wildfires. The nature of fighting wildfires requires a level of scientific understanding of fires, and the technical skill to fight them, that is (1) not required for water-based hazards and (2) not readily available in non-firefighting organizations. In the case of the Saskatchewan wildfires, such knowledge requirements took extra time for LOs to acquire and to disseminate to CAF units, constraining the amount of time available to provide an overview of the general provincial response. Frontline troops, in turn, faced a significant learning curve in understanding fire behaviour and how to effectively mitigate fire growth. Unlike the skills required to fight floods and ameliorate hurricane damage, soldiers could not leverage their existing skills when fighting wildfires. Furthermore, wildfires were the only hazard with a historically well-defined profession dedicated to fighting it, which meant that professional firefighting units used entrenched internal organizational models not always intuitive to external organizations, including CAF, which in turn meant that the 2015 Saskatchewan event was the only one where some civilian officials were ambivalent about CAF as an ideal disaster response partner.

The unique impact of wildfires challenge the all-hazards approach supported in much of the EM literature and practice in Canada and the US. Disaster research has identified an all-hazards EMO as a potential solution to collaboration problems during response since the mid-twentieth century (Rosow 1955; Williams 1956; Form and Nosow 1958), and essentially all Canadian provinces and the US federal government, the respective levels of government primarily responsible for emergency management in each country, have developed "all-hazards models" over the last century and into the current one (Scanlon 1995; Knowles 2011; Lindsay 2014).

Both the academic and practitioner arguments for an all-hazards approach hinge on the organizational efficiency of developing relatively few strategies, and allocating relatively few resources, to protect a system's vulnerabilities (regardless of the hazard that exploits the vulnerabilities), versus developing a multitude of potentially overlapping and redundant strategies – and attempting to find an array of mitigating resources – for each potential hazard. The general acceptance of such an approach has been undermined by hazard-specific organizations in the case of security (i.e., when the "hazards" are directly human-caused), but less so in the case of natural hazards.

The organizational need to anticipate hazard-specific effects in the case of wildfires is not necessarily a rejection of the all-hazards approach. Indeed, part of the reason LO-driven information flow was less robust in Saskatchewan was due to the wildfire management branch playing an almost EMO-type role in managing the wildfires. An argument could be made that if the actual EMO had taken more of a lead such information flow could have been improved. Wildfires may have had a hazard-specific impact not only because fires themselves are unique (the specialized skills required to fight fires cannot be ignored), but because wildfires were the only natural hazard that, once developed into a disaster, was in part managed by a hazard-specific agency. The hazard-specific agencies monitoring the hazard development of floods and Hurricane Igor were not involved in managing the actual disaster responses. In other words, the challenge that wildfires pose to the all-hazards model may not be that wildfires should be moved outside of such models, but that wildfires should be more fully integrated into such models. Either way, pending changes to the implementation of EM policy as it relates to fighting wildfires, both EM academics and practitioners should anticipate hazard-specific effects on interorganizational collaboration when a disaster is caused by wildfires.

## The Role of Institutions and Weak Federal Response Capacity

The oft-repeated refrain in the public administration literature that "institutions matter" requires nuancing in the context of domestic disaster response. Not only did informal networks play a large role, but formal rules – such as the RFA confirmation process – were often secondary to informal and functional processes playing out among civilian and military organizations closer to the disaster zone. Furthermore, differences in the particular bureaucratic configurations of each province and Joint Task Force (JTF) played practically no role in the nature of interorganizational collaboration; *presence*, *quality*, and *barrier* levels rose and fell largely independent of jurisdictional context. Indeed, the nature of civilian and military organizations themselves (i.e., the informal links organizations had fostered among themselves; the degree to which organizations were hampered by benign incapacity; the conceptual differences between organizations; and so on), more so than the unique institutional contexts faced by organizations across the different events, largely explained the nature of domestic disaster response, from the decision-making processes to the implementation of response actions.

The "nature of organizations themselves" can, of course, be argued to be a function of the broader institutional context. However, the institutional context of Canadian governance writ large does not explain the intricate connections that CAF – a federal resource – built up with provincial and municipal officials, up to and including the point of CAF LOs self-identifying as provincial or local resources. Conceptual differences between civilian and military organizations are, in turn, more convincingly explained through the functions of each organization than through the particular institutional contexts that provide their respective mandates; a "provincial military" or "municipal militia" would likely not have solved the conceptual differences that existed between the "federal military" and its civilian disaster response partners. Only municipal benign incapacity is intuitively explained through an institutional approach given that (1) the level of government with the most resources, the federal government, is not directly linked to municipalities in Canada's constitution; and (2) the federal and provincial governments are the constitutionally mandated actors in Canadian public policy and administration, which provides them with far greater decision-making and revenue-generating capacity than municipal governments. The benign incapacity of the RCAF, however, is more easily explained by the limited political appeal of spending on CAF than by any particular institutional dynamic.

The glaring exception to "nuancing" the effects of institutions was the role of civilian federal actors during disaster response. The constitutional divide between the federal level of government and the level most immediately and directly affected by the disasters (municipalities), combined with provincial governments as the main disaster response coordinators (due in part to Canada's vast geography, but also due to the responsibilities constitutionally allocated to provinces), meant that federal organizations close to the disaster zones, namely Public Safety Canada and the Public Health Agency of Canada (PHAC) were largely "information monitors" that would provide situational awareness to the Government Operations Centre (GOC) in Ottawa.

CAF officers and non-federal civilian officials were highly aware that the inability of their regional federal counterparts to initiate the use of significant resources or dictate response tasks was not due to a lack of such resources or individual capability, but to a lack of authority and mandate. When federal organizations did touch base with municipal officials, they established a connection only because "they wanted to brief Ottawa" (Interview 2, 2017). At the provincial level, officials mainly engaged federal representatives when such representatives asked for updates and did not think to involve the regional federal actors in the details of disaster response formulation; "the Feds were vacated to the background" (Interview 17, 2018).[4] While CAF and provincial EMOs were the main actors at the strategic level of disaster response coordination, and CAF and municipalities were the main actors at the tactical level of implementing disaster response actions, the civilian federal government – once the RFA had been cleared – was institutionally constrained from substantially affecting the response phase.

While Canada's weak federal response capacity stands in stark contrast to the American and, albeit to a lesser degree, the Australian domestic disaster response systems,[5] such weakness cannot be characterized as problematic for effective emergency management. Not a single CAF officer or EM official could identify a specific example where solutions to disaster response problems, be they at the strategic problem-solving level or at the tactical resource-heavy level, would have more easily been reached and/or improved with greater federal presence and capacity at decision-making tables or the front line. Furthermore, individuals across military and civilian lines preferred Canada's decentralized model to the American one where the US Federal

Emergency Management Agency (FEMA) plays a substantial role during disaster response (Interviews 2, 6, 8, 9, 14, 15, and 17, 2017). One official did have a contrasting perspective and noted that, despite the general success of the CAF–province–municipality "response team," they were concerned about national coordinating capacity should a disaster have multi-provincial or national impacts; they believed that not only is a "Canadian FEMA" warranted, but that EM at the federal level should have regulatory teeth to enforce compliance with EM regulations and standards (Interview 4, 2017). This official worked at the federal level and was based in Ottawa throughout all the disasters assessed here.

To summarize, federal response capacity in Canada (1) is weak compared to the other levels of government and CAF; (2) is limited by institutional, not resource, constraints; and (3) does not adversely affect the ability of the disaster response system to formulate and implement actions during the response phase. The academic implications of such claims allow the hypothesis that EM systems in federal countries may not require "top-heavy" disaster response systems where the federal level of government plays a strong role during the response phase. Future research can identify comparable disaster events across Canada, the US, and Australia to assess the degree to which robust versus minor federal response capacity influences effective disaster response. Overall, this book affirms the existing Canadian EM and public administration literature that describes the federal government as a limited disaster response actor compared to the provinces (Henstra and Sancton 2002; McEntire and Lindsay 2012; Hale 2013; Juillet and Koji 2013).

The implications for the civil–military relations literature is that CAF, a federally funded and mandated organization, was perceived by all the actors in the disaster response system as a resource "belonging to all," and was not directly associated with the federal government. Unlike combat missions overseas, there may be something about domestic response to natural hazards that has the potential to dissociate an armed force from the power of the state, and to link such a force with a civic service that transcends any particular level of government.

The implications that a weak federal response capacity has for practitioners are twofold. First, provincial and municipal emergency managers should include templates for federal government-specific situation reports in their emergency plans (as the primary concern of regional federal organizations is collecting information for Ottawa), and they should be careful about building reliance on non-CAF federal decision-making or resources into such plans. Second, emergency managers with skills specialized for the response phase of emergency management may be better utilized at the provincial and/or municipal levels. Individuals with interests in or skills specialized in the other phases, particularly mitigation and recovery, may be effective at the federal level.

*The Limits of Integration: A Chink in the Collaborative Framework's Armour*

Since the academic literature has not specifically addressed the effectiveness of the CAF–civilian relationship during domestic response to natural disasters prior to this work, an original collaborative framework was developed based on the EM literature on multi-organizational disaster response, the public administration literature on the implementation of EM policy, the civil–military relations literature on Canada's military and joint civilian–military projects, and CAF's domestic operations doctrine. Components important to effective multi-organizational disaster response, public policy implementation, and civil–military relations were organized into the three concepts of the presence and quality of, and barriers to, interorganizational collaboration, and indicators were developed for each component.

The research process affirmed the legitimacy of the framework as an assessment tool as both CAF officers and civilian officials generally concurred with the desirability-level designated to each component by the framework. The "assessment perspectives" embedded in the framework and those of the research participants aligned because components that would increase or decrease the presence or quality of, or barriers to, interorganizational collaboration in the framework were understood as legitimate increases or decreases by officers and officials alike. For example, both the framework and participants concurred that the presence of interorganizational collaboration was increased through higher levels of information sharing, flexibility, collective conflict resolution, and so on; that the quality of such collaboration was increased through more coordination, higher levels of satisfaction, trust, and so on; and that barriers to such collaboration occurred through lack of resources, conceptual difference, empire-building, and so on.

The alignment of the framework and practitioner assessments was not, however, universal. Two indicators for the component of integration were linked in the framework to the quality of interorganizational collaboration, while most participants viewed such indicators as either irrelevant or, indeed, detrimental to such collaboration during disaster response. The civil–military relations literature and CAF domestic operations doctrine informed the indicators for the component of integration, which – at its most robust – sees not only liaison mechanisms between organizations or joint decision-making at a strategic level, but the interoperability of members from different organizations (i.e., "integrated activities") at the front line, and the interoperability of planning processes. Both of these indicators were low across the events studied, resulting in a low level of integration and only a moderately high quality of interorganizational collaboration overall. Participants, however, never characterized the military–civilian separation on the front line as problematic to effective disaster response. Front-line response actions delivered by *either* a military or civilian organizational unit were described purely in practical terms, and unlike the other "undesirable" indicators in the framework (e.g., lack of sharing emergency plans, unwillingness to work with another organization again), non-interoperability at the front line was discussed by participants from a neutral or positive perspective.

The main theme distilled from the participants' perspectives on front-line non-interoperability was the (unanticipated) functionality of such non-interoperability. In Saskatchewan, the ongoing confusion about the civilian Incident Command System (ICS) and the military chain of command, combined with discrepancies in civilian versus military firefighting ability, meant that professional firefighters were allocated to "hotter" spots than the soldiers standing in as firefighters (Interview 18, 2017). Across Manitoba and Alberta, the maintenance of the military aspect most appreciated by civilian authorities – self-sufficiency – required that CAF units run their own sandbagging lines and fight their own floods so that CAF commanders could reallocate resources speedily wherever was most necessary without worrying about accommodating civilians "under their command" (Interviews 15 and 16, 2017). Similarly, CAF engineers in Newfoundland had specific tools, standard operating procedures, terms, and general tradecraft, the efficiency of which would have been inhibited through attempting to absorb a civilian crew (Interview 1, 2017).

The framework's continued development requires greater sensitivity for detecting the way in which even CAF's own domestic operations doctrine may be influenced by CAF's central mandate and method, which is the defence of Canada and the legal application of violence, respectively. Front-line interoperability as the epitome of integration makes sense in a combat context as all front-line organizations are military in nature, with similar internal processes (the chain of command), tools (weapons), functions (the application of force), and goals (vanquishing "the enemy"). Integration was not only desirable for Canadian soldiers conducting attacks alongside other NATO troops, but was also feasible. This book demonstrates that the natural disaster context – where response partners are not professional soldiers, and the enemies are flame and water – warrants a division of labour between civilians and troops at the front line. Note, however, that removing interoperability as an indicator for integration, and therefore its importance to the quality of interorganizational collaboration, would not have drastically altered the overall results. Other indicators for integration – common measures of effectiveness and, albeit to lesser degree, common situational understanding – were also low across all events. The quality of interorganizational collaboration would have been higher, but room for improvement would have remained.

The academic implication of the above is twofold. First, scholars researching the effectiveness of non-combat civilian–military projects should pay careful attention to the context that informs a military's doctrine. Such doctrine can be useful in informing a framework's accuracy (the other CAF doctrine-informed indicators for coordination and integration were aligned with research participants' assessments of what was (in)effective disaster response), but can also inconspicuously introduce elements that do not automatically transfer from combat to non-combat contexts. The second implication, which is equally relevant to practitioners, is that the functionality of interorganizational collaboration has limits. As is demonstrated in chapter 2, the EM literature largely embraces collaboration as a good without limits; the more collaboration the better the disaster response. However, collaboration as an end in itself, versus a means to achieve a specific and desirable disaster response action (such as increasing the amount of information to improve strategic decision-making across organizations; increasing the trust that allows civilian authorities to leave some response actions to other organizations; increasing the degree to which different organizations agree on disaster scope in order to efficiently use resources; and so on), can lead to ineffective outcomes. None of the CAF officers and civilian officials involved in leading disaster responses that spanned four provinces, three hazards, and two JTFs, saw the achievement of specific and desirable ends through maximizing collaboration on the front line.

## Competing Bureaus: The Dark Side of Disaster Response

While including front-line interoperability was a weakness in the collaborative framework, adding the concept of "barriers to interorganizational collaboration," and defining that concept in part through an empire-building component informed by the public administration literature on competition among public sector organizations, was a strength. Such competition is predicted by economic theories of the bureau, where organizational anti-collaboration is characterized as logical, functional, and unexceptional when public sector organizations – or bureaus – have to draw from the same well of resources. For example, if withholding information or resources from other organizations could help organization X use that information or resources to receive credit – and thereby gain prestige, influence, and ultimately, funding – then, according to economic theories of the bureau, organization X would withhold information and/or resources even to the detriment of a socially desired outcome, such as effective disaster response. Such a prediction was incorporated into the framework in the event that the nature of the disaster events did not only spark varying levels of the desirable indicators of the components under the presence and quality of collaboration, but also the undesirable indicators of empire-building.

The incorporation of an economic theory of the bureau did pay off in the results. While empire-building was virtually non-existent between CAF and civilian organizations, and between levels of government, empire-building was high within the same levels of government. Federal organizations kept their information chains discrete from each other (which allowed for clarity around which organizations deserved credit for information arriving at the GOC); provincial EMOs and health emergency management organizations kept both information and resources from each other; and in some cases, municipal officials prioritized their own neighbourhoods for disaster response service delivery. In each case, organizations and/or individuals only exhibited empire-building or undesirable competitive behaviour when they were drawing from the same well of resources.

The implication of the above for theorists is twofold. First, economic theories of behaviour within public administration can add another empirical example of organizational behaviour that affirms the role of competitive behaviour in the public sector. Second, the "collaboration niche" within the EM literature should not prioritize the ideal requirements of organizational behaviour during disaster response without acknowledging the perverse incentives that may exist for organizations during such responses. If the framework had been informed only by the EM literature, it would have ended after the assessment of the quality of interorganizational collaboration. While such an assessment would not have yielded uniformly "desirable" outcomes (as is demonstrated through some of the indicators for coordination and integration not occurring), it also would have not have captured behaviour explicitly "negative" (such as the indicators for empire-building that did occur). Emergency management scholars should incorporate general theories of public sector behaviour developed in public administration to a greater degree rather than assume the unique nature of the disaster context will nullify or inhibit such behaviour.

The implication for practitioners is also twofold. At an individual level, emergency managers should be aware of the resources that support the various organizations they work with during disaster responses, and they should consider whether any organizations may be incentivized to compete with each other during such responses. Reasons provided for not including organizations at decision-making tables or distribution lists should be explicitly linked to the effectiveness of a specific disaster response action versus a more general claim that organization X simply does not need to be present or informed. At a strategic level, emergency planners or, indeed, emergency management policy-makers, should carefully weigh the pros of function-specific organizations and avoiding one organization's monopoly over a policy area versus the cons of intragovernmental competition when contemplating the creation of separate agencies involved in EM within the same level of government. Public organizations involved in EM are not immune from expanding their mandates to increase their potential resource share. Just as the conflation of national security and domestic law enforcement led to the US's Central Intelligence Agency (CIA) and its Federal Bureau of Investigation (FBI) keeping crucial information and resources from each other in the lead-up to 9/11 (National Commission on Terrorist Attacks upon the United States 2004), so too can the mandate-creep of two organizations in Canadian emergency management yield unhealthy competition.

### RFAs and End States: Swamps of Conceptual Difference

Indicators for components across the presence and quality of interorganizational collaboration suffered due to different civilian versus military understandings of the scope of the Request for Assistance (RFA) and the value of end states. Conceptual differences regarding the RFA meant that officials (especially municipal officials) used time and resources planning for CAF response actions that the military could not fulfil, and that CAF officers spent time and resources educating civilians on the scope of the RFA and explaining why some civilian requests could not be granted. Emergency managers in and out of uniform lamented that resources spent on negotiating the nature of the

RFA could have been better spent on more speedily developing implementable response actions.

Conceptual differences regarding end states were even more problematic as not only did such differences compound different understandings of the legitimate level of CAF involvement (i.e., the scope of the RFA), but they also meant that different measures of effectiveness were used across military and civilian lines. Ultimately, non-manipulative influence, coordination, and integration were not "maximized" due to different military versus civilian conceptions of the RFA and disaster end states. Conceptual difference, more than any other barrier, adversely affected CAF–civilian collaboration during disaster response.

The main theoretical implication of the above is that differences in perceiving reality played a bigger role than material resources in determining levels of interorganizational collaboration. The resource constraints faced by the RCAF and municipalities were ultimately overcome by, in the case of the former, efficient resource reallocation by the other CAF branches and, in the case of the latter, resource support from the other levels of government and CAF. Indeed, as limited local resources are by definition characteristic of disasters, benign incapacity outside of the RCAF was viewed more as the reason for – versus a weakness of – the disaster response. Furthermore, officers and officials often noted, but rarely lamented, resource constraints; such constraints were disaster problems to be solved. In contrast, they lamented when other organizations did not respect – or over-respected, depending on the participant's perspective – the limits of the RFA or the development of a clear end state. In other words, while tolerance for limited resources was generally high across the board, and therefore was not detrimental to interorganizational collaboration, tolerance for other organizations viewing important parts of the disaster response in different ways was low. Emergency management, public administration, and civil–military relations research that focuses on multi-organizational endeavours would therefore do well to include a constructivist bent in their frameworks. As one senior CAF officer, responsible for assessing lessons learned from military–civilian disaster responses, stressed: "Perception is reality" (Interview 5, 2017).

The main implication for practitioners, in turn, is that much damage can be done simply by not understanding the perspective of one's partner organizations. A disaster response partner may arrive with significant resources, but if that partner's conception of the legitimate use of such resources is not understood then their ability to effectively deploy those resources will be constrained. Time and energy would have been far more efficiently and effectively spent by all response organizations if CAF's role (1) as an aid to civilian authorities (i.e., CAF was not going to take charge); (2) as including no security mandate (i.e., CAF was not going to enforce the law); and (3) as only available for the disaster's response phase (i.e., CAF needed to redeploy the moment municipalities could hold their own) had been understood and internalized by civilian authorities prior to the disaster events. On the flip side, much time and energy would have been more efficiently and effectively spent had CAF internalized prior to deployment the understanding that civilian authorities (1) viewed the military as "the cavalry" come to plug gaps wherever necessary and as therefore unencumbered by what, from the civilian point of view, often seemed like arbitrary constraints; and (2) could not afford to think of disaster relief as constrained to the response phase as they would be living in and supporting the disaster-struck communities long after the hazard had disappeared. While neither the military nor the civilian side would, given their unique mandates, change their points of view on the aforementioned, awareness of each other's points of view would render conceptual differences into disagreements. As the former are implicit differences in perceiving reality and the latter are explicit differences in what can or cannot be done, significant planning time and energy will not be spent on the latter.

The good news for practitioners here is that EM policy-makers can make substantial improvements to disaster response without requesting extra funds. "More resources" is the inevitable call from all policy spheres, which creates a funding competition that emergency management – given its perennially low political salience, with its relevance punctuated but not sustained by disaster events – is unlikely to win. The benign incapacity of municipal EM and the RCAF are problems for effective disaster response to be sure, but more funding for either will be more difficult to achieve than using existing resources to diminish the level of conceptual difference that exists between military and civilian organizations on the matter of the RFA and end states. A basic education program for civilian officials on the legal scope of the RFA and why CAF has to redeploy after the response, as well as incorporating civilian perspectives into CAF's domestic operations exercises, would be a start.

### *"Shared Responsibility" Affirmed: Joint Civil–Military Decision-Making*

The salience of interorganizational collaboration during the CAF–civilian responses studied here, combined with the moderately high quality of and low barriers to such collaboration, demonstrate that

154  Boots on the Ground

during disaster response in Canada, civilian and military disaster response organizations share responsibilities at both the plan formulation and implementation levels. As described when assessing the components of information sharing and flexibility under the *presence* concept, and the components of trust and satisfaction under the *quality* concept, CAF was not only involved strategically through joint decision-making processes in provincial EOCs, but civilian organizations were also comfortable with allocating a formulating role to soldiers. The civil–military relations paradigm, popular in and out of academia throughout the mid-to-late twentieth century and still prescribed by niches within the contemporary literature, that sees bureaucrats as "planners" and troops as "doers" in civilian–military endeavours was not the reality on the ground. While civilian authorities had the ultimate legal say on troop deployment and, indeed, made the request that allowed troop deployment in the first place, once troops hit the ground the response formulation and implementation were shared across CAF and its civilian counterparts.[6]

As predicted by the civil–military relations "shared responsibility" niche, military involvement at the formulation level was not only the functional reality in practice, but was more effective than attempting to constrain the military to acting as "doers." More than half of the desirable indicators that occurred under the presence and quality of civilian–military collaboration occurred at the strategic level, and the civilian reliance on the battle rhythm set by CAF – which broke seemingly unmanageable challenges into discrete tasks – occurred through senior official exposure to senior CAF officers during the response phase. The theoretical implication here is that neither effectiveness nor liberal democratic norms need be undermined by including military personnel on the strategy team. To the contrary, such inclusion improved effectiveness and allowed CAF officers to express their adherence to the legal superiority of civil authorities. The implication for practitioners, in turn, is that front-line non-interoperability should not bubble up to the strategic level, where plan formulation through joint civilian–military brainstorming needs to be fostered and maintained.

**Policy Recommendations**

The rationale for this research enquiry was to address a puzzle hitherto uninvestigated in the academic literature, but also to improve emergency management in Canada. This book, as befits public policy and administration research, was therefore written for both the inherent value in expanding knowledge and the practical value of improving real-world outcomes. While the book's contribution to knowledge – valuable to both academics and practitioners – is discussed in the general themes above, improvement to emergency management in Canada requires system-level suggestions on what governments can do to improve outcomes in the emergency management domain.

The overall results and their general themes inform three policy recommendations that are elaborated below. The first two recommendations avoid the calls for "more research" and/or "more funding" that so often conclude analyses of this sort and instead highlight policy directions that can be implemented largely without the need for more resources. Indeed, while there are areas where further study and extra funds would likely improve outcomes (and three are noted in the third recommendation), a meta-theme that emerged from this work is that barriers to effective disaster response are often related to communication flow through networks and different conceptions of reality, which are both dynamics that can be mitigated without acquiring additional human resources or physical assets.

### Foster and Expand Emergency Management Networks

A focus on the response phase of EM is not blind to the other phases, but rather illuminates what parts of preparedness were effective. Across all the disasters examined here, the response phase saw CAF officers and civilian officials rely on connections established with other individuals during routine operations. Links created between individuals who would be involved in disaster response decision-making and action development were the parts of preparedness – more so than formal emergency plans, lessons learned from exercises, or the readiness of physical equipment – that were immediately used and the most relied upon when floods, fires, and wind became disastrous. The first policy recommendation is therefore aimed at all disaster response organizations across all levels of government and suggests that fostering and expanding an organization's links to other organizations relevant to disaster response is as important to effective disaster response as ensuring internal readiness and capacity.

Specific ways this policy recommendation can be implemented includes maximizing EOC "cross-contamination" during non-disaster times and disaster lead-up. Almost half of all the potential indicators for the presence of interorganizational collaboration occurred because CAF had LOs embedded into provincial EOCs whenever EOCs were used for regularized status updates on the "hazard landscape" (e.g., monthly meetings), temporarily "stood up" to monitor a specific hazard, or activated full-time to manage an ongoing disaster. Provincial EOCs are not, however, the only EOCs in each province. The relevant CAF JTF, health emergency management organizations, the regional PHAC and Public Safety Canada organizations, and many municipalities may all have EOCs. If such EOCs were conceptualized not as discrete hubs for each organization (with the organization's individuals as spokes attached to each hub), but rather as plants that require bees from all disaster response organizations in order to be pollinated and thrive, then the high levels of military–civilian collaboration sparked by the CAF LO presence in the provincial EOCs could be expanded across the disaster response system.

Another way of implementing this policy recommendation would be through requiring a disaster response organization to include individuals from a variety of other disaster response organizations in the facilitation and/or observation of exercises, and in the lessons learned processes (or after action reviews) that follow exercises or actual events. While there was some civilian inclusion in some of CAF's exercises leading up to the events studied here, civilian and military participation in each other's exercises is not institutionalized and is viewed as something "nice to do" rather than something that is fundamentally necessary. Lessons learned processes, in turn, were largely conducted with no external presence. Such processes have to be conducted in an environment where individuals can speak freely, up to and including criticizing the actions of other organizations, but requiring some portions of after action reviews to include external input would have allowed the opportunity for different organizations to discuss their different perceptions of concepts such as disaster end states or RFAs, which in turn may have reduced the adverse impacts such differences in perceptions had on subsequent events.

Whatever form this policy recommendation takes in specific organizations, and whether it is implemented as a "small-p policy" that informs disaster response organizations' best practice or a "big-P Policy" that requires organizations with disaster response mandates to "cross-pollinate," it will address at least three of the general themes discussed above. Fostering and maintaining EM networks will not only buttress the informal networks so essential to effective disaster response and allow for opportunities to reduce the conceptual differences so detrimental to disaster response, but such networks may – through the links individuals create that transcend their particular organization – help to offset some of the empire-building seen within the same levels of government.

## Maintain Decentralized Emergency Management in Canada

Canada does not have a federal agency operationally empowered to coordinate disaster response in the same way that other public safety-related agencies at the federal level, such as the Royal Canadian Mounted Police or the Canadian Security Intelligence Service, are operationally empowered to pursue their mandates. The US, however, does have an agency comparable to its FBI and CIA when it comes to disaster response: the Federal Emergency Management Agency (FEMA). The idea of a "Canadian FEMA," modelled on this US federal agency, was brought up by almost half of the research participants, but only one – a civilian official working for the federal government – believed that such a model would improve disaster response in Canada. This book affirms the view of most participants that the vulnerabilities that do exist in Canada's disaster response system would not be ameliorated through importing the "top-heavy" US model and thereby creating an operationally empowered emergency management agency at the federal level in Canada.

Canada's decentralized system that allocates the bulk of disaster response coordination to the provinces and municipalities, with support from regional CAF JTFs, meant that (1) the provincial EMOs and municipal leaders fostered direct links with regional and local CAF units without going through the extra organizational layer of the federal government, and (2) the individuals geographically and institutionally close to the disasters were the ones making decisions about how best to manage them. None of the disasters saw organizations absent during routine times swoop in to take control during non-routine times, which meant the crucial formal and informal networks that existed in the disaster zones were empowered to take action. Furthermore, Ottawa's involvement in regional and local disaster response decision-making would likely have decreased the solidarity and collegiality near and in the disaster zones as non-federal officials expressed annoyance at the Government of Canada's constant requests for updated information, never mind had the capital identified tasks or directives for regional or local officials to follow.

The oft-repeated phrase in the EM literature that "all disasters are local" can be nuanced to "all disaster responses are local." No participant questioned the role of the federal government in mitigation through research and policies or leveraging the Government of Canada's fiscal capacity to support provinces and municipalities in recovery. The emphasis was on *decentralized processes during disaster response*, and indeed, just as provincial EMOs and regional CAF JTFs were empowered to jointly develop disaster response coordination strategies in each affected province (leading to troop readiness and therefore expeditious deployment), so too were mayors and platoon commanders empowered with any information required to take response actions on the ground as they saw fit (avoiding delays that would occur through relaying all potential actions through the provincial level). Except in cases where conceptual difference existed on whether a response action fell outside the RFA or after a disaster's end state, the military doctrine of "mission command," which sees leadership closest to a problem initiate and act on a solution, was emblematic of disaster response in Canada.

"Mission command," which is essentially an operational phrase for what in institutional language would be "decentralization," addresses at least three of the general themes discussed above. Not only is mission command tailor-made to work within the low federal response capacity that is the reality of Canada's EM system, but it also allows military and civilian organizations to respond together not as prescribed through a conventional civil–military relations doctrine that separates civilians into "planners" and soldiers into "doers," nor through a popular niche in the EM literature that maximizes even non-functional interorganizational collaboration during tasks on the front line. Canada's decentralized system (1) bestows strategic responsibility for response actions based on institutional and geographical proximity to the event, not on organization type (i.e., senior officers and provincial officials work together on strategy); and (2) allows front-line officers and civilians to lead their respective units as they find most functional, which is through non-integration on the front line.

This policy recommendation largely advocates for the status quo and cautions against importing centralized models from other countries without assessing the functional behaviour incentivized by Canada's existing decentralized model. While EM policy-making at the federal level may be effective across the other EM phases, from research that assesses how climate change may increase the frequency and/or impact of different hazards (mitigation), to funding for provinces that require more extensive emergency plans and exercises of their municipalities (preparedness), to regulating insurance schemes that incentivize development outside hazard zones (mitigation and recovery), the results here indicate that policies that change the decentralized nature of Canadian EM during the response should be avoided.

### Enhance RCAF, Municipal, and All-Hazards EM Capacity

Canadian disaster response can benefit from fostering and expanding EM networks and affirming the functional dynamics inherent in Canada's decentralized response system, neither of which require significant funds for extra human resources, physical assets, or research. However, as the reality of public administration means that demands for public services are rarely satiated by the existing supply, potential improvement in services – including disaster response services – is possible through increased resources. Such extra resources for the system that responds to the type of disasters that require multi-governmental and military support should take the form of more funds to the RCAF and municipalities, and more research on how wildfires can be incorporated into the all-hazards model of emergency management.

The only area of CAF support that was deemed problematic by both CAF officers and civilian officials was the lack of sufficient air assets to transport equipment, relief supplies, troops, officials, and citizens (i.e., evacuees). RCAF is not only challenged by being responsible for the second biggest landmass overseen by air forces across the world (a challenge mitigated by the Army through its multitude of reservist units that form a part of regular communities across the country and by the Navy through its constrained focus on coastal areas), but also by the fact that the air branch is the only CAF branch responsible for a routine emergency service separate from "the integrity of the state" – that is, all air search and rescue across the country. These two challenges meant that air assets were not supplied even close to the demands required by the disaster needs on the ground across the events studied here. Indeed, even though informal civilian–military networks ensured troop readiness across the events, the speed of relief supplies, evacuations, and arrival of civilian officials near disaster sites could likely all have been improved through the availability of more air assets.[7]

The other area that would improve disaster response in Canada through the receipt of more funding is municipal EM capacity. Many municipal levels of government did not understand the RFA (and hence spent valuable time putting together requests that either made little sense to CAF officers, such as requests for specific assets versus for effects, or could not be granted, such as requests to conduct law enforcement) and/or could not contribute fully to the development of disaster response actions as they were net recipients of such actions (and hence not "full players" in disaster response). Both issues could be ameliorated in part through funds provided specifically to bolster municipal EM capacity. It is one thing for a province to mandate that municipalities develop emergency response plans, but it is quite another to fund full-time emergency coordinators that review RFAs, develop exercises that test plans, participate in the provincial EOC, liaise with local CAF reservist units, and so on. Such extra human resource capacity – not to mention the type of funding that would allow municipalities to run annual public awareness campaigns on disaster preparedness, incentivize development outside of hazard zones, and build resilient infrastructure – would allow municipal officials to not only understand the processes that will define provincial and CAF response actions, but also allow them to make decisions about and contribute to such actions.

The final area that warrants extra resources is research on incorporating wildfire response into existing all-hazards models. Such models have been largely effective and adopted across Canada and the US; replacing them with ones where hazard-specific agencies take on the coordination mantel is neither realistic nor desirable. However, all-hazards models have not fully integrated wildfire management into their response processes, which can lead to adverse outcomes. Wildfire response saw a provincial wildfire management branch take on significant parts of the coordination role, including liaison with CAF, which meant that CAF officers on the ground did not have the province-wide awareness enjoyed by officers in the other events. The rich history and professional identity of firefighters – versus the lack of a cohesive history and identity for those who fight other natural hazards – also meant that officials felt ambiguous about CAF support, adversely affecting the sense of interorganizational solidarity and collegiality that supports the maintenance of informal networks; and that established internal firefighter structures clashed with CAF's own established internal structures.

The all-hazards model worked well across the cases where the all-hazards EMO held a monopoly on the ultimate coordination decisions and where the internal structures of the organizations that joined CAF on the front line were informal and ahistorical compared to the military's. Future research therefore must assess how an all-hazards model can be applied when a set of specific organizational processes and professional attributes have developed around – and maintained control of coordinating – a specific hazard, especially if those processes and attributes have to occur alongside another set of specific organizational processes and professional attributes, such as those found in a professional military.

## Conclusion

The general themes that emerged from this work include the crucial role of informal networks, the hazard-specific impact of wildfires, the effects of weak federal disaster response capacity, front-line integration as unnecessary collaboration, intragovernmental competition for resources, the Request for Assistance and "end states" as swamps of conceptual difference, and the importance of military–civilian "sharing" of the decision-making role. Such themes, in turn, informed the development of three policy recommendations to further improve Canada's disaster response system, namely the fostering and expansion of emergency management networks; the maintenance of Canada's decentralized disaster response system; and additional funds and research to, respectively, enhance RCAF and municipal emergency management capacity, and determine how the all-hazards emergency management model can more effectively incorporate wildfire management during the response phase.

# Conclusion

Emergency management's focus on interorganizational collaboration as a key variable in an array of desirable disaster response outcomes, civil–military relations' identification of shared responsibility as key to successful joint civilian–military endeavours, and public administration's recent emphasis on engaging all relevant stakeholders when addressing complex policy problems were all affirmed by analysing disaster response in Canada. However, some of the assessments embedded in each of these relevant literatures need to be nuanced. The emergency management and public administration literatures need to recognize that interorganizational collaboration should not be an end in itself in the public sector, with fully interoperable planning processes and personnel on the front end of service delivery especially liable to frustrating some desirable outcomes. In a similar vein, the civil–military relations literature needs to appreciate the limits of applying the same high levels of military–military integration in a combat context to civilian–military endeavours in a non-combat context.

Future research can nuance the contribution of this work. Natural disasters tend to be viewed as "acts of God" and thus generate more cooperative responses and less blame-avoidance and finger-pointing compared to a hazard or threat that is composed – or perceived to be composed – of human-caused elements. Security-related or industrial disasters are therefore potentially less likely to see leaders of disaster response organizations characterize the actions of other such organizations in a positive light compared to when organizations respond together during natural hazards. While the all-hazards approach is affirmed here, with levels of collaboration similar across a variety of natural hazards, the way behaviour changes during non-natural hazards may have an effect on the nature of CAF–civilian response to security or industrial events. The generalizability of CAF–civilian collaboration can therefore be tested through future research that assesses such collaboration during responses to non-natural hazards.

It should be noted, however, that security- and industrial-focused research may encounter a central methodological challenge avoided by this work on natural hazards; that is, no historical precedent apart from the 1917 Halifax harbour explosion exists for the scope of military contribution seen in large scale natural disaster response. The isolated impact of domestic security or industrial events – compared to floods, wildfires, hurricanes, ice storms, and so on – means that the military contribution required to meaningfully assess CAF–civilian response effectiveness in the first place may not occur.

The generalizability of this work can potentially be expanded through future research that includes more front-line disaster responders as research participants. While this book includes three "strategy versus tactics" sections – under the assessment of flexibility in chapter 4, integration in chapter 5, and conceptual difference in chapter 6 – where differences between high-level decision-makers and front-line implementers are discussed, the data used to inform the nature of such differences came from archival analysis and a participant pool made up largely of non–front-line officials and officers. Such a pool was required for the nature of this particular enquiry because the research puzzle could only be addressed through information from individuals with insight into the overall disaster responses, but future research could directly analyse the perspectives of front-line responders in order to buttress empirical claims about tactics as well as strategy.

Finally, studying joint CAF–civilian disaster response can be layered through assessing whether gender and/or other group characteristics affect the degree to which civilian officials and CAF officers collaborate. This research distinguished between individuals based purely on the nature of their disaster response organization (civilian versus military, and municipal versus provincial versus federal); such an approach fit the research puzzle that was orientated to the nature of CAF-civilian collaboration at an organizational, not a sociological, level. Future research, however, may be able to discern links between the unique backgrounds and lived experiences of civilian officials and CAF officers, how they collaborate with each other, and hence how their organizations collaborate during disaster response. For example, numerous studies have shown that gender and ethnicity can affect how and the degree to which a variety of hazards and activities are perceived as risks (Gustafson 1998; Leikas et al. 2007; Sloan, Chepke, and Davis 2013; Morioka 2014; Herrero-Fernández et al. 2016; Novie et al. 2018). Future research can assess how and whether such varying levels of group-dependent risk perception affects the way individual officials and officers conceptualize the impact that hazards will have on communities, and whether such potentially varying conceptualizations in turn affect the levels of CAF–civilian collaboration when such hazards become disasters.

Whatever the specific direction, the extension of this research would further help illuminate a corner of public policy and administration in Canada that rarely enjoys the spotlight. Despite the regular occurrence of joint CAF–provincial–municipal disaster response, where uniformed soldiers work shoulder-to-shoulder with civilian leaders to shield communities from harm, academic analysis and media reporting rarely assess the nature of such joint response in the discussion of emergency management or the public policy process writ large.

David Johnston, the twenty-eighth governor general of Canada, noted the following:

> [T]he history of warfare and the history of the military are often usefully seen as distinct. In the settlement and growth of Canada from the 1600s on, it was the military that performed the critical function of surveyors, engineers, architects, surgeons, and builders of towns and forts, ships and boats, canals and bridges that together were the vital links we used to forge a peaceful and prosperous new society. (Johnston 2016)

While such deep involvement of the military in civilian projects is neither a reality nor a necessity in contemporary Canada, soldiers do indeed become surveyors, engineers, architects, surgeons, and builders in Canadian communities when disasters strike. During those most difficult of times, when jurisdictions and citizens have exhausted their resources, soldiers integrate into civilian society to form a unified disaster response team. Such integration is not only to be commended, but warrants expanded and continued analysis.

# Appendix A: Basic Incident Command System (ICS) Organizational Chart

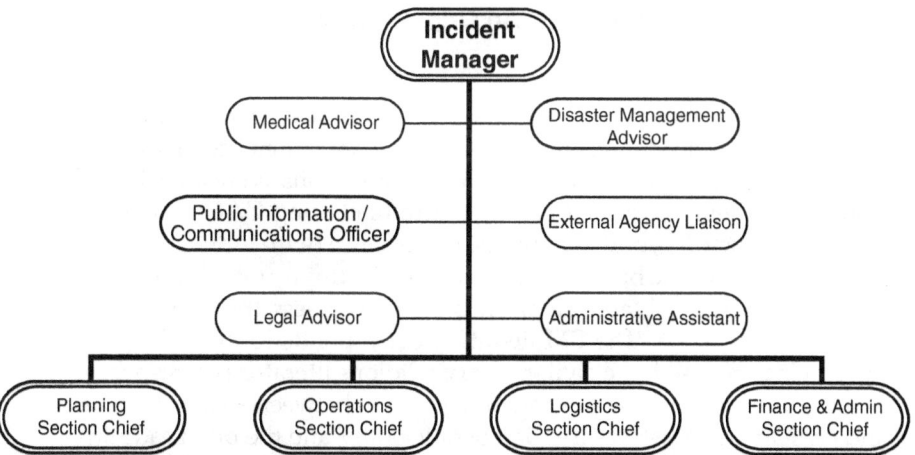

# Appendix B: Methodology

**Fusing the Three Approaches: An Original Collaborative Framework**

The approaches that inform this study's collaborative framework originate from three academic literatures: interorganizational collaboration from the emergency management (EM) literature, shared responsibility from the civil–military relations literature (including CAF's domestic operations doctrine), and new public governance (with input from economic theories of the bureau) from the public administration literature. All three bodies of literature are necessary to assess the phenomena studied in this book. The EM literature corresponds to the policy domain being analysed, the civil–military relations literature corresponds to the unique nature of analysing the dynamics between an armed force and civilian organizations in a democratic polity, and the public administration literature corresponds to the implementation of public policy.

Table 7. Three Approaches from Three Literatures

|  | Interorganizational Collaboration | New Public Governance (NPG) | Shared Responsibility |
|---|---|---|---|
| Main discipline | Emergency management | Public administration | Civil–military relations |
| Critical response to: | Hierarchical systems (e.g., the Incident Command System) | Previous public administration paradigms (i.e., traditional public administration) | Civilian policy formulation only (e.g., Huntington's model) |
| Prescribes collaboration between: | Disaster response organizations | Policy problem stakeholders | Civilian and military leaders |

Furthermore, the approaches from each of the three literatures can be applied to the same set of actors relevant to this research.

Table 8. The Relevant Actors

|  | Disaster Response Organizations (Interorganizational Collaboration) | Policy Problem Stakeholders (NPG) | Civilian and Military Leaders (Shared Responsibility) |
|---|---|---|---|
| Relevant actors | CAF; federal, provincial, and municipal governments | CAF; federal, provincial, and municipal governments | CAF; federal, provincial, and municipal governments |

Each approach yielded components (at times overlapping) that are essential to effective outcomes during multi-organizational disaster response, joint civil–military projects, and policy implementation. These components are information sharing, non-manipulative influence, flexibility, support, collective conflict resolution, coordination, integration, satisfaction, trust, and processes to deal with conceptual difference. As these traits are all "desirable" in that they indicate the type of collaboration that leads to effectiveness, the economic theory of the bureau literature was searched for key traits that inhibit collaboration; they are empire-building, manipulation, and distrust. Benign incapacity is a trait hinted at by the EM literature and CAF's domestic operations doctrine as a non-competitive trait that inhibits collaboration, and so can be added to the latter list. The effectiveness of CAF's contribution to domestic disaster response can now be assessed by observing the degree to which these components are present in the system. As effectiveness is broken down into three research questions and three corresponding concepts, the first step is to allocate the components to one of the three concepts.

Table 9. The Concepts and Their Components

| Research Question | Concept |
|---|---|
| 1. What is the role of CAF in contemporary Canadian disaster response? [descriptive] | The *Presence* of Interorganizational Collaboration<br>- Information sharing<br>- Non-manipulative influence<br>- Flexibility<br>- Support<br>- Collective conflict resolution |

(*Continued*)

| Research Question | Concept |
|---|---|
| 2. How effective is the civilian–military relationship during disaster response? [evaluative] | The *Quality* of Interorganizational Collaboration<br>- Coordination<br>- Integration<br>- Satisfaction<br>- Trust |
| 3. In what ways might the military contribution to Canadian disaster response be improved? [normative] | The *Barriers* to Interorganizational Collaboration<br>- Empire-building<br>- Manipulation<br>- Distrust<br>- Benign incapacity<br>- Conceptual difference |

The *presence* of interorganizational collaboration components were placed under the umbrella of the first research question because their observation describes the existence of interorganizational collaboration. As the specific indicators for these components will show, they are less dynamic than the other traits, and while they can exist on a 0–1 scale, they do allow a "check-in-the-box" measurement of existence. Furthermore, even a high level of all of these traits would not by itself demonstrate a highly collaborative system; these traits merely indicate that a baseline of collaboration is present, and that the foundation for deeper collaboration exists. The components under the concept of the *quality* of interorganizational collaboration, on the other hand, are more complex and require more elaboration from individuals within the disaster response system to determine the nature of their existence in the system. Finally, the presence of traits under *barriers* to interorganizational collaboration allows normative suggestions to be made on what should be done to improve gaps in the collaborative aspects of the disaster response system. Indicators of each component are provided below.

### The Presence of Interorganizational Collaboration

*Research Question 1: What is the role of the CAF in contemporary Canadian disaster response? [descriptive]*

COMPONENTS
- *Information sharing* (formal *and* informal)
    - **Indicators:** Distribution of emergency response plans; presence in Emergency Operations Centre (EOC); presence at

decision-making table; maintenance of communication during response actions; distribution of new data as event unfolds; description of organizational strengths and vulnerabilities; access to a variety of individuals from another organization (i.e., elected and non-elected officials in civilian government; higher and lower ranks in military).
- ○ **Interview Questions:** Military (M): 4, 5, 9 | Civilian (C): 4, 5, 8
- *Non-manipulative influence* (organizations are not autonomous; decision-making is participative)
  - ○ **Indicators:** Decision-making capacity at decision-making table; capability to change actions of others in a non-manipulative way; lack of institutional constraints to helping disaster response partners.
  - ○ **Interview Questions:** M: 7, 8, 9, 10 | C: 7, 9, 10
- *Flexibility*
  - ○ **Indicator:** Willingness to reach a decision not originally anticipated by the organization.
  - ○ **Interview Questions:** M/C: 9, 10
- *Support* (physical and non-physical resources are pooled to aid collective goals)
  - ○ **Indicator:** Any form of resource augmentation (i.e., technological, human, funds, intelligence) of another's efforts.
  - ○ **Interview Questions:** M: 5, 6 | C: 7
- *Collective conflict resolution* (organizations work together to solve conflicts)
  - ○ **Indicators:** Conflicts explicitly brought to a designated collective table; input of other organizations or a third party sought.[1]
  - ○ **Interview Questions:** M/C: 9, 10, 12

## The Quality of Interorganizational Collaboration

*Research Question 2: How effective is the civilian–military relationship during disaster response? [evaluative]*

COMPONENTS
- *Coordination*
  - ○ **Indicators:** A common vision of the end state; general agreement on the actions necessary to achieve the end state; an understanding that collaboration is more effective and efficient than non-collaboration; strong liaison mechanisms, including embedding staff where feasible; understanding of the impact of

one organization's activities on other organizations; consultation prior to taking action, when feasible.
    - **Interview Questions:** M/C: 1, 6, 14
- *Integration*
    - **Indicators:** A common purpose; a common situational understanding; a common or interoperable planning process; a unity of effort with integrated activities designed to achieve the common objective; common measures of effectiveness.
    - **Interview Questions:** M/C: 6, 15
- *Satisfaction*
    - **Indicator:** Expressed approval of another organization's actions.
    - **Interview Questions:** M/C: 11, 16
- *Trust*
    - **Indicators:** Willingness to work with another organization again; willingness to have an individual from another organization dictate actions (e.g., use of resources) for the organization; willingness to share data.
    - **Interview Questions:** M/C: 4, 11, 16
- *Selflessness*
    - **Indicator:** Providing any dimension of collaboration above with the knowledge that such actions could harm the organization's symbolic resources (i.e., adversely impact its reputation, morale, etc.).
    - **Interview Question:** M/C: 26

## The Barriers to Interorganizational Collaboration

*Research Question 3: In what ways might the military contribution to Canadian disaster response be improved? [normative]*

COMPONENTS
- *Empire-building*
    - **Indicator:** Withholding information that could benefit joint goals to enhance the power of one organization.
    - **Indicator Questions:** C: 4
- *Manipulation*
    - **Indicator:** Changing or attempting to change the behaviour of others in a veiled manner in order achieve an organization's goals over joint goals.
    - **Interview Question:** M/C: 27

- *Distrust*
  - **Indicators:** Unwillingness to share information; unwillingness to leave response actions to others; preference for working alone.
  - **Interview Question:** M/C: 4, 6, 19, 20, 21, 22, 23
- *Benign incapacity*
  - **Indicator:** Non-collaboration due to a lack of resources (i.e., technical expertise, equipment, funds, etc.)
  - **Interview Questions:** M/C: 20A
- *Conceptual difference*
  - **Indicator:** Incompatible definitions of the problem and/or the solution (i.e., different characterizations of the scope of disaster; different ideas of who/what is to blame; different political views on who is responsible for disaster response; different appreciations for the best type of organizational model to use in disaster response; varying knowledge-levels/interpretations of emergency management best practices).
  - **Interview Questions:** M/C: 6, 17, 18, 19, 20

**Linking Selected Cases to the Theoretical Framework**

The four selected case studies are used to assess levels of interorganizational collaboration across this study's three research questions. The answers to each of the three research questions can confirm one of two hypotheses:

1 Descriptive Hypothesis: CAF plays [/does not play] a collaborative role in contemporary Canadian disaster response.
   a. Levels of information sharing are high [low]
   b. Non-manipulative influence is evident [not evident]
   c. Flexibility is high [low]
   d. Support is high [low]
   e. Collective conflict resolution is evident [not evident]
2 Evaluative Hypothesis: The civilian–military relationship during disaster response is effective [/not effective].
   a. Levels of coordination are high [low]
   b. Levels of integration are high [low]
   c. Levels of satisfaction are high [low]
   d. Levels of trust are high [low]
3 Normative: The military contribution to Canadian disaster response should [/should not] be improved.
   a. Levels of empire building are high [low]
   b. Levels of manipulation are high [low]

c. Levels of distrust are high [low]
   d. Levels of benign incapacity are high [low]
   e. Conceptual difference is evident [not evident]

While the hypotheses above are set up as binaries for methodological clarity, the qualitative approach employed here to assess for each indicator of each component means that the answer to each question can land somewhere along a spectrum. For example, the potential outcome after interviewing a participant involved in a specific event may read as follows: "CAF played a *moderately* collaborative role in event X because three out of the five descriptive components were indicated; CAF was *largely* ineffective in event X because only one out of the four evaluative components were indicated; and CAF's contribution to events like X can *generally* be improved because three out of five normative components were indicated."

The interview outcome can be nuanced through adverbs such as "moderately," "largely," and "generally" because the qualitative approach employed allows (1) participant answers that exist along a spectrum, and (2) an assessment of a participant's responses made by the researcher. This approach is preferred to one that would yield "either/or" answers as it is unlikely that the complexity of disaster response sits at either 0 or 1 on any variable (i.e., the civilian–military relationship is likely not purely (in)effective or the disaster response is likely not un-improvable). The descriptive, evaluative, and normative results of this research approach will be more nuanced than an either/or result, which in turn will allow for an exploration of *levels* within the presence and quality of, and barriers to, interorganizational collaboration, and – crucial to scholars and practitioners alike – how such levels may be increased or decreased.[2]

The final assessment of a specific event will require the collation of all participant responses for that event. The assessment of each component depends not only on how an individual participant answers a question that indicates a component (i.e., emergency plans were widely shared, indicating the descriptive component of information sharing), but on the degree to which participants from the same event and engaged in the same response actions agree when indicating components (i.e., civilian and CAF leaders who provide different answers on whether emergency plans were widely shared indicate not only a lack of information sharing by some participants, but other components as well, such as the normative component of distrust). This is one of the reasons why the theoretical framework explicitly includes evaluative and normative components. Half of the participants from the same event may indicate a descriptive component such as information sharing or support, which does not by itself say much about the quality of interorganizational collaboration, since half of the participants did not indicate such components. Such a result, however, does demonstrate lack of integration, the presence of conceptual difference, or perhaps benign incapacity, indicating that interorganizational collaboration during the disaster response experienced shortcomings.[3]

The comparison of responses requires that participants are grouped into their respective disaster events. Their answers to each question are assessed individually as important information about the disaster response, but are also compared to answers to the same question provided by the other participants in their group.[4] The results section delineates between and discusses these two assessment processes for each event, especially as wildly varying components of collaboration indicated by participants involved in the same event indicate a high level of conceptual difference, which in itself is a problem for genuine interorganizational collaboration. The comparison of participant responses is a way of not relying only on the perceptions of emergency managers/leaders in assessing for levels of collaboration. The comparison of different perspectives on the same event/response actions allows a level of analysis that does not take one individual's perceptions as a perfect mirror of reality. Rather, perceptions from different individuals that "sit" in different disaster response organizations are assessed to build a picture of the overall response and the level of interorganizational collaboration that occurred.

Once an assessment is made of the presence and quality of, and the barriers to, interorganizational collaboration during each event, a cross-case comparison can be made across the different events to discern whether any links exist between levels of collaboration and a specific hazard and/or jurisdiction. Participant responses also informed a results discussion on why indicators of the components of interorganizational collaboration occurred or did not occur. Furthermore, any links between specific EM practices that emerged from the interviews – not already included in the theoretical framework's indicators – and levels of interorganizational collaboration will be discussed.

## Semi-Structured Interviews

### THE USE OF INTERVIEWS

In-depth interviews with government officials is an essential and established tool used to analyse the public policy process (Jiwani and Krawchenko 2014). Indeed, the Canadian EM domain has mainly been studied through interviewing individuals involved in the EM policy formulation process (Hale 2013; Juillet & Koji 2013; Grieve and Turnbull 2013). There are challenges, however, when using interviews to research the experience of and response to disaster events themselves, which have been documented from disaster studies' formal inception up to and throughout the twenty-first century.[5] Not only does the impromptu nature of disasters frustrate the anticipation of events that may fit a study's requirements (Cisin and Clark 1962), but interviews "in the field" (i.e., during a disaster) are fraught with difficulties, whether they are the ethical considerations of interviewing disaster survivors as they experience trauma and emergency managers as they attempt to control a crisis situation, or the data accuracy concerns of asking people for their perception of an phenomenon while they are under high stress and/or functioning under the narrow focus that both disaster survival and emergency management on the ground can require (Stallings 2007). Nevertheless, the continuing research interest in human experience during disaster, including direct assessments of organizational response to disasters, means that interviews are well-established method to glean information about disaster events, albeit often after the event has occurred.[6]

Interviewing participants after disaster events allows researchers to (1) find relevant cases in history (versus chasing them as they occur); (2) be informed by analysis already performed on cases; and (3) collect more reflective perspectives that are not preoccupied with either meeting or providing for the basic needs of others. Such criteria are essential to a research project oriented towards assessing for detailed and complex indicators that inform the degree to which a concept such as interorganizational collaboration occurred. Not only do the indicators for evaluative and normative components like integration, satisfaction, and empire-building require a reflective, bird's eye view of response actions, but – as their presence may ebb and flow depending on how the disaster response develops, which itself depends in part on the unpredictable evolution of the hazard – they can only be assessed fully once the response phase is over. While some descriptive indicators, such as the presence of different organizations in EOCs, could be assessed during a disaster, research resources would limit the ability to track personnel flow across the many EOCs active during significant disasters.[7] Furthermore, as discussed below, this research requires high-level participants who would scarcely have time for interviews during an event, the consequences of which will largely fall on their shoulders. The reasons for conducting interviews after events are therefore methodological, ethical, and practical in nature.

Problems faced when selecting disaster research participants have also been noted since the field's infancy. A disaster affects lives and property within a geographical space in varying ways, causes varying psychosocial and financial effects on others not within physical reach of the hazard, and disrupts peoples' regular lives up to and including physical relocation. The upshot is that disaster conditions "make it difficult to define, locate and reach the universe to be sampled" (Killian 1956). This problem is avoided by this research project through the type of participants required to answer the research questions. Participants in this project need to have been either high-level civilian public sector managers or high-ranking military officers during the event. Only decision-makers in charge of or intricately involved with an organization's overall disaster response strategy would have the knowledge to answer questions that probe for the indicators of each component. Front-line soldiers and municipal workers would not be able to indicate whether the lieutenant-general or provincial minister directing their actions was present at multi-organizational EOCs, shared emergency plans with each other, (dis)trusted other organizations, allowed resources to be used or directed by other organizations, and so on. This research's population is, in other words, relatively small, as it contains only those individuals who were involved in steering an organization's disaster response strategy.

Therefore, selected research participants were (1) individuals who were far enough removed from the hazard to minimize the likelihood of becoming disaster survivors themselves (i.e., they were "sent in" to manage the event, which endowed them with the sense that they were there to deal with the disaster, as opposed to the disaster purely being something adverse happening to them); and (2) individuals who, given their stature in directing public organizations during disaster response, are publicly identifiable through media and government reports on the event. The research participant selection process therefore avoided the methodological, ethical, and practical challenges many disaster researchers face when researching the experience of disaster survivors themselves.

High-level civilian managers at each level of government and senior CAF officers were identified as point-of-first-contact subjects using media and government reports on each of the selected disaster events. From there a snowball technique was used to identify other relevant and important players.[8] Each potential participant received a formal research invitation Email that detailed the scope and value of the research, as well as the nature of their potential contribution. Individuals willing to participate signed consent forms and were provided with information on the steps taken to protect their identities.[9] For ethical as well as methodological reasons, no mention was made to any participants of the roles, titles, positions, mandates, or specific government organizations (below a departmental/JTF level) that may have applied to the other participants.

While the snowball recruitment technique meant nonprobability sampling, methodologically robust answers to the research questions still required three essential criteria of the research participant selection process in that the process needed to yield (1) multiple perspectives to be queried about specific response actions to minimize individual bias; (2) only individuals who performed in a role senior enough to speak to the research's indicators; and (3) both civilian and military perspectives that were evenly represented. The first criterion encourages a sample at least larger than one participant per event, the second constrains the size of the population from which to sample, and the third constrains the selection process through requiring evenly distributed participation from two discrete groups.[10]

Consideration of all three requirements above yielded a sample of twenty participants. Case studies of each disaster event and its response actions were informed by four participants: two officers and two civilians who were at or near the top of their organization during the response. Another four participants who were involved in all the case studies' disaster events at or near the top of their organization were recruited to provide a direct comparative assessment of the disaster responses.[11] The lack of a larger sample is not seen as a problem because (1) as a nonprobability sampling, qualitative, and case study driven research project, severe limits to generalizability already exist; and (2) selecting more participants would have violated the second criterion above by including individuals who could not adequately speak to this research's indicators. Indeed, the selection process allowed for each individual within the group of twenty research participants to have had significant decision-making capacity – and hence insight into – disaster responses. While respecting the confidentiality of the participants prohibits the publication of their exact roles and organizational affiliations, it can be stated that civilian participants were directors of emergency management organizations mandated to coordinate the disaster response and/or ministers responsible for the department overseeing the disaster response. Military participants were experienced commissioned officers responsible for leading a CAF contribution to the disaster response. These participants not only had a high-level overview of the disaster responses studied, but they were the ones setting and adjusting the response strategy as the disaster developed. They could speak to each of the identified indicators to assess levels of interorganizational collaboration.

The interview rubrics that guided the assessment of the participants were semi-structured with a set list of questions that allowed for subjects to discuss themes they found important. This approach builds on an established interview tradition that seeks to minimize interviewer bias through careful development of a set list of questions while providing room for research participants to identify important elements the researcher may not have anticipated (DiCicco-Bloom and Crabtree 2006; Stake 2010; Qu and Dumay 2011).

Archival analysis of each case study informed the interviews so that the researcher could engage participants using a thorough knowledge of the disaster response in question. Participants were first asked about the immediate lead-up to the disaster response and the processes used to initiate CAF's contribution. These questions probed for differences in perceptions of the evolution of hazard development and the information flow among organizations as the hazard became a disaster. Once the disaster lead-up had been discussed, the participants were asked about the roles they, their organization, and other organizations played throughout the response. These two sections yielded a document that probed for each indicator of each component, producing a picture of the presence and quality of, and barriers to, interorganizational collaboration during disaster response.

### Linking Interviews with the Theoretical Framework

The interview questions probe for each indicator of each component. While the questions generally probe for indicators in the order that they appear in the theoretical framework (from descriptive to evaluative to normative indicators), most questions probe indicators of more than one component, which means that some evaluative and/or normative components are probed by earlier questions as well as later ones. Indeed, the interview rubrics were constructed to allow an intuitive flow for participants and not to align perfectly with the order of components and their indicators as they appear in the theoretical framework.

The interview rubric for civilian emergency managers/CAF leaders is provided below. The component(s) being probed follows each question. Probes break down evenly across the descriptive, evaluative, and normative concepts; components informing each concept are probed fifteen times (i.e., the interview rubric probes for components a total of forty-five times).

Section 1. Formal Deployment Process and Disaster Lead-Up

| Interview Question | Components Probed |
|---|---|
| 1. From your vantage point in organization X, can you explain the official process needed to initiate CAF deployment during a domestic disaster? | Coordination; Conceptual difference |
| 2. During event X, did your experience more or less reflect the official process required to initiate a CAF deployment during a disaster? If not, how did it differ? | Coordination; Conceptual difference |
| 3. When hazard X occurred, the federal government announced that CAF would be deployed on date X. From your vantage point, when was the idea that CAF might be used first shared? Who would you say initiated the idea? | Information sharing |
| 4. Did you have direct contact with anyone from CAF before the federal government announced that CAF would deploy?<br>• If yes, were emergency response plans shared? Did CAF discuss areas where its response capacity would be useful versus less useful, robust versus less robust? Did CAF suggest how the Request for Assistance should be put together and what should be included? Were you engaged with a variety of CAF members or one main link into the organization, such as a liaison officer? | Information sharing; Empire-building |
| 5. Does your organization engage with CAF during "calm" times when no hazards are occurring or imminent?<br>• If yes, with whom? Do you talk about potential response strategies or weak areas should certain disasters occur? Do you feel that CAF is comfortable sharing ideas, strategies, and plans with your organization?<br>• If no, where do you get most of your information about CAF response capacity and strategy? When do you get that information? | Information sharing; Empire-building; (Dis)Trust |
| 6. During hazard X, did the disaster response plans of CAF dovetail with your disaster response plans? For example, did it appear that CAF plans specifically included elements on how to work with your organization/level of government?<br>• Is there ever pushback within your organization on involving CAF in disaster response? If yes, do departments within government, or elected versus non-elected officials, differ in their support for involving CAF? Is there any difference between the levels of support for CAF involvement in municipal compared to provincial governments? In general, what is the level of support within your organization for involving CAF in disaster response? | Coordination; Integration; (Dis)Trust; Conceptual difference |

Section 2. Interorganizational Collaboration during Disaster Response

| Question | Component (Not) Indicated |
|---|---|
| 7. Because CAF legally deploys as *aids* to civilian power, the task force requires some type of civilian oversight of its actions throughout the response. During hazard X, were the details of the CAF operation guided by the federal, provincial, or municipal governments, or a combination of them? Did the task force have discretion from the federal government to provide aid where provincial or municipal governments required it?<br>• Was the aid that was provided by CAF negligible, important, or essential to the response? What form did the aid come in (technological, human, financial, or intelligence)? | Non-manipulative influence; Support |
| 8. Once deployed, did the task force communicate mainly with the provincial government, or with the municipal governments where support was taking place, or both?<br>• Regarding the communication that your level of government received directly from CAF during the response, did it occur in the Emergency Operations Centre? In other words, did representatives from your organization and CAF both have seats in the EOC? Did you receive new data from CAF members on the ground as the response unfolded fairly quickly (within twenty-four hours), or only after a couple of days? In general, do you feel like communication was maintained between your organization and CAF – to the degree that communication was relevant – during the response? | Information sharing; (Dis)Trust |
| 9. Before response actions in location X, did the task force sit down and discuss procedures with your organization?<br>• If yes, did your organization establish its priorities followed by how the task force could help meet those needs, or did the task force have its own suggestions for what should be done first? In other words, did the government focus more on *what* should be done while the task force focused more on *how* to do it? Were there ever disagreements between CAF and civilian leaders on what or how something should be done? If yes, how were they solved? Was CAF happy to pursue a response action they had not initially suggested? Was your organization? Do you feel like your organization had genuine decision-making influence/capacity at the decision-making table?<br>• If the task force did not sit down with your organization before response actions, where did clearance for that action come from? Did any conflict arise from a lack of direct involvement from your organization? If so, how was that conflict approached/handled? Did it linger throughout the response or was it satisfactorily addressed? | Non-manipulative influence; Flexibility; Collective conflict resolution |

*(Continued)*

| Question | Component (Not) Indicated |
|---|---|
| 10. During hazard X, who had the final say on taking a specific response action? Were there clear lines of authority according to level of government, CAF, or agency type, or was it more of a consensus building process? | Non-manipulative influence; Flexibility; Collective conflict resolution |
| 11. Looking back at the disaster response, was your organization satisfied with how CAF worked with the municipal and/or provincial governments? Would your organization be happy to work with CAF again under similar circumstances? If not, what got in the way of a satisfactory relationship? | Satisfaction |
| 12. Could your organization ever sense conflict between CAF and the provincial, municipal, or federal levels of government on what actions to take and how to take them? If yes, who had the ultimate say? Were those decisions respected and acted upon? | Collective conflict resolution |
| 13. Was working with CAF easier than working with the other levels of government? If yes, what made it easier? | Satisfaction |
| 14. During the response phase, would you say that your organization and CAF shared common criteria to define what the end of the response would look like? In other words, was there a shared vision for the disaster response "end state"?<br>• If yes, was there general agreement on the specific response actions necessary to achieve the end state? Was there an understanding that collaborating with organizational partners is part of an effective and efficient response? Did strong liaison mechanisms exist between your organization and CAF (such as embedding staff in each other's organizations where feasible)? Was there an appreciation of the impact one organization's activities may have on other organizations? Did consultation occur – when feasible – before specific response actions took place on the ground? | Coordination; Conceptual difference |
| 15. In a similar vein, would you say that your organization and CAF shared a common sense of purpose throughout the response? In other words, did you share a common understanding of the situation as it unfolded (such as why a disaster was occurring and how the hazard would evolve)? Did you plan together throughout the response? Were response actions integrated to achieve common objectives? Did your organization and CAF generally agree on whether a response action was successful or not? | Integration; Conceptual difference |
| 16. Was there any type of direct follow-up with your organization after CAF's support work was completed? For example, were there joint reviews on how well the disaster response was coordinated? Did your organization's internal reviews include assessment on how well CAF worked with your organization/level of government? If so, what was the general assessment? | Satisfaction |

| | |
|---|---|
| 17. During the response phase, were you and/or your organization willing to have CAF help dictate your organization's actions and/or use of resources? Would you and/or your organization be willing to work with CAF again in the future? Given your overall experience during the disaster response, is there anything that could be changed to improve how CAF supports civilian authorities during disasters? | (Dis)Trust |
| 18. Did CAF make a positive difference in how the disaster was managed? If yes, was the difference substantial? Compared to other organizations, what allows CAF to make a positive difference? | Satisfaction; Support |
| 19. In general, does your organization feel that there is an important role for CAF to play during disaster response? | Satisfaction; Support |
| 20. Overall, would you say that collaboration between different organizations is important to effective disaster response, or is one competent organization in control more important?<br>• If collaboration is more important, does your organization/level of government have enough human, technological, equipment, and budgetary, and other such resources to work effectively with other levels of government/CAF? In other words, during the disaster response, could your organization "keep up" with the other organizations' capabilities? | Empire-building; Benign incapacity; (Dis)trust |
| 21. Does your organization need other levels of government and/or CAF during disaster response, or if your organization had enough resources would it manage the response better on its own? | Empire-building |
| 22. Overall, would you and/or your organization characterize other levels of government and/or CAF as "partners in" or "barriers to" effective disaster response? | Empire-building |
| 23. How would your organization characterize its overall goal during the disaster response: for the organization to work as well as it can with other levels of government/CAF, or for the organization to complete its own specific tasks as well as it can? | Empire-building |
| 24. Did your organization see response to hazard X as an opportunity for the organization, or was it more of a burden? In other words, is disaster response understood as a chance for your organization to show its capabilities, or is it understood as tasks that contain risks to positive public perception, the budget, and so on? | Benign incapacity |
| 25. During the disaster response, what was the most important: getting the other levels of government/CAF to do what your organization had identified as important response actions, or getting everyone to agree on the appropriate response actions? Were there ever times where achieving an important response action required withholding some information from other organizations? | Manipulation |

# Appendix C: Archival Analysis Sources for Each Case Study

Note: The full citation for each source appears in the bibliography.

### 2010 Hurricane Igor | Newfoundland and Labrador

Andowski and Auld 2010; Associated Press 2010; Canadian Press 2010; CBC News 2010a, 2010b, 2010c, 2011b, 2011c; CBC News in Review 2010; CTV News 2010; DND and CAF n.d.b; Environment and Climate Change Canada n.d.a; Fire and Emergency Services – Newfoundland and Labrador 2011; Government of Canada 2010; Gutro 2010; Ibrahim 2014; Iype and Bouzane 2010; National Oceanic and Atmospheric Administration 2010; Newell 2010; Newfoundland and Labrador Statistics Agency 2010; Pasch and Kimberlain 2011; Postmedia News 2010

### 2011 Assiniboine Flood | Manitoba

Canadian Press 2013a; CBC News 2011a, 2011d; CTV News 2011; DND and CAF 2011; Environment and Climate Change Canada n.d.b; Government of Manitoba n.d.a, n.d.b, 2011a, 2011b, 2011c, 2013; Public Safety Canada 2013, 2015; Redekop 2011; Rural Municipality of Macdonald 2011; Rural Municipality of St. Francois Xavier 2011; Stunden Bower 2011

### 2013 Multi-River Flood | Alberta

Calgary Herald 2013, 2014a, 2014b; Canadian Army n.d.a, 2013; Canadian Press 2013b; CBC News 2013; Davison 2013; Environment and Climate Change Canada n.d.c; Garson and Canadian Press 2013; Gerson 2013; Government of Alberta n.d., 2013a, 2013b, 2014; Ogrodnik 2013

## 2015 Wildfires | Saskatchewan

Amiro et al. 2011; Botha 2015; Burton et al. 2008; Canadian Army n.d.b; Canadian Interagency Forest Fire Centre n.d.; Canadian Press 2015a, 2015b; CBC News 2015; Charlton 2015; Eyre 2015; Giles 2015; Government of Saskatchewan n.d; Johnson, Miyanishi, and Bridge 2001; La Ronge EMS 2015; Mortillaro 2015; Natural Resources Canada n.d.a, n.d.b; Parisien et al. 2006; Reid 2015; Stocks et al. 2002; *The Wildfire Act*

# Appendix D: Collated Results

|  | NF 2010 | | MB 2011 | | AB 2013 | | SK 2015 | |
|---|---|---|---|---|---|---|---|---|
|  | Civilian | Military | Civilian | Military | Civilian | Military | Civilian | Military |
| *The Presence of Interorganizational Collaboration* | | | | | | | | |
| **1. What is the role of CAF in contemporary Canadian disaster response? [descriptive concept]** | | | | | | | | |
| 1.1 Information Sharing | | | | | | | | |
| 1.1.a Distribution of emergency response plans | 2 | 2 | 2 | 2 | 2 | 2 | 2 | 2 |
| 1.1.b Presence in Emergency Operations Centre | 2 | 2 | 2 | 2 | 2 | 2 | 2 | 0 |
| 1.1.c Presence at decision-making table | 2 | 2 | 2 | 2 | 2 | 2 | 2 | 0 |
| 1.1.d Maintenance of communication during response actions | 2 | 2 | 2 | 2 | 2 | 2 | 2 | 2 |
| 1.1.e Distribution of new data as event unfolds | 2 | 2 | 2 | 2 | 2 | 2 | 0 | 2 |
| 1.1.f Description of organizational strengths and vulnerabilities | 2 | 2 | 2 | 2 | 2 | 2 | 2 | 2 |
| 1.1.g Access to a variety of individuals from another organization | 2 | 2 | 2 | 2 | 2 | 2 | 2 | 0 |
| **Result: 104/112 = 93% (HIGH)** | | | | | | | | |
| 1.2 Non-Manipulative Influence | | | | | | | | |
| 1.2.a Decision-making capacity at decision-making tables | 2 | 2 | 2 | 2 | 2 | 2 | 2 | 1 |
| 1.2.b Capability to change actions of others in non-manipulative way | 2 | 2 | 2 | 2 | 2 | 2 | 2 | 1 |
| 1.2.c Lack of institutional constraints to helping disaster response partners | 0 | 0 | 0 | 0 | 0 | 0 | 0 | 0 |
| **Result: 30/48 = 63% (MODERATE)** | | | | | | | | |
| 1.3 Flexibility | | | | | | | | |
| 1.3.a Willingness to reach a decision not originally anticipated by organization | 2 | 2 | 2 | 2 | 2 | 2 | 2 | 1 |
| **Result: 15/16 = 94% (HIGH)** | | | | | | | | |

1.4 Support

| | | | | | |
|---|---|---|---|---|---|
| 1.4.a Any form of resource augmentation (e.g., tech, human, funds, intel) of another's efforts | 1.75 | 1.75 | 1.75 | 1.75 | 1.75 |
| **Result: 7/8 = 88% (HIGH)** | | | | | |
| 1.5 Collective Conflict Resolution | | | | | |
| 1.5.a Conflicts explicitly brought to a designated collective table; input of other organizations or a third party sought | 2 | 2 | 2 | 2 | 2 |
| **Result: 16/16 = 100% (HIGH)** | | | | | |
| **Total Presence: 86% (HIGH)** | | | | | |
| *The Quality of Interorganizational Collaboration* | | | | | |
| **2. How effective is the civilian–military relationship during disaster response? [evaluative concept]** | | | | | |
| 2.1 Coordination | | | | | |
| 2.1.a Common vision of the end state | 0 | 0 | 0 | 0 | 0 |
| 2.1.b General agreement on the actions necessary to achieve the end state | 0 | 0 | 0 | 0 | 0 |
| 2.1.c Understanding that collaboration is more effective and efficient | 2 | 2 | 2 | 2 | 2 |
| 2.1.d Strong liaison mechanisms, including embedding staff where feasible | 2 | 2 | 2 | 2 | 2 |
| 2.1.e Understanding the impact one organization's activities may have on other organizations | 2 | 2 | 2 | 2 | 2 |
| 2.1.f Consultation prior to taking action, when feasible | 2 | 2 | 2 | 2 | 2 |
| **Result: 64/96 = 67% (MODERATE)** | | | | | |
| 2.2 Integration | | | | | |
| 2.2.a Common purpose | 2 | 2 | 2 | 2 | 2 |
| 2.2.b Common situational understanding | 1 | 1 | 1 | 1 | 1 |
| 2.2.c Common or interoperable planning process | 0 | 0 | 0 | 0 | 0 |
| 2.2.d Unity of effort with integrated activities designed to achieve the common objective | 0.5 | 0.5 | 0.5 | 0.5 | 0.5 |
| 2.2.e Common measures of effectiveness | 0 | 0 | 0 | 0 | 0 |
| **Result: 28/80 = 35% (LOW)** | | | | | |

*(Continued)*

|  | NF 2010 | | MB 2011 | | AB 2013 | | SK 2015 | |
|---|---|---|---|---|---|---|---|---|
|  | Civilian | Military | Civilian | Military | Civilian | Military | Civilian | Military |
| 2.3 Satisfaction | | | | | | | | |
| 2.3.a Expressed approval of another organization's actions | 2 | 2 | 2 | 2 | 2 | 2 | 2 | 2 |
| **Result: 16/16 = 100% (HIGH)** | | | | | | | | |
| 2.4 Trust | | | | | | | | |
| 2.4.a Willingness to work with another organization again | 2 | 2 | 2 | 2 | 2 | 2 | 2 | 2 |
| 2.4.b Willingness to have an individual from another organization dictate actions (e.g., use of resources) for the organization | 2 | 2 | 2 | 2 | 2 | 2 | 2 | 2 |
| 2.4.c Willingness to share data | 2 | 2 | 2 | 2 | 2 | 2 | 2 | 2 |
| **Result: 48/48 = 100% (HIGH)** | | | | | | | | |
| **Total Quality: 76% (MODERATELY HIGH)** | | | | | | | | |
| *The Barriers to Interorganizational Collaboration* | | | | | | | | |
| **3. In what ways might the military contribution to Canadian disaster response be improved?** [normative concept] | | | | | | | | |
| 3.1 Empire-Building | | | | | | | | |
| 3.1.a Withholding information that could benefit joint goals to enhance the power of one organization | 2 | 0 | 2 | 0 | 2 | 0 | 2 | 0 |
| 3.1.b Withholding resources that could benefit joint goals to enhance the power of one organization | 2 | 0 | 2 | 0 | 2 | 0 | 2 | 0 |
| 3.1.c Conceptualizing response actions as tied to the goals of one organization | 2 | 0 | 2 | 0 | 2 | 0 | 2 | 0 |
| **Result: 24/48 = 50% (MODERATE)** | | | | | | | | |
| 3.2 Manipulation | | | | | | | | |
| 3.2.a Changing or attempting to change the behaviour of others in a veiled manner in order achieve an organization's goals over joint goals | 0 | 0 | 0 | 0 | 0 | 0 | 0 | 0 |
| **Result: 0/16 = 0% (LOW)** | | | | | | | | |

**3.3 Distrust**
3.3.a Unwillingness to share information — 0 0 0 0 0 0 0 0
3.3.b Unwillingness to leave response actions to others — 0 0 0 0 0 0 0 0
3.3.c Preference for working alone — 0 0 0 0 0 0 0 0
**Result: 0/48 = 0% (LOW)**

**3.4 Benign Incapacity**
3.4.a Non-collaboration due to a lack of resources (i.e., technical expertise, equipment, funds) — 1 1 1 1 1 1 1 1

**Result: 8/16 = 50% (MODERATE)**

**3.5 Conceptual Difference**
3.5.a Incompatible definitions of the problem and/or the solution (i.e., different characterization of the scope of disaster) — 2 2 2 2 2 2 2 2
3.5.b Different ideas of who/what is to blame — 0 0 0 0 0 0 0 0
3.5.c Different views on who can take a response action — 2 2 2 2 2 2 2 2
3.5.d Different appreciations for the best type of organizational model to use in disaster response — 0 0 0 0 0 0 0 0
3.5.e Varying knowledge levels/interpretations of emergency management best practices — 1 1 1 1 1 1 1 1

**Result: 44/80 = 55% (MODERATE)**
**Total Barriers: 31% (LOW)**

Note:

With four participants – two civilian and two military – selected for each disaster response, each indicator could occur a maximum of four times for each event. The maximum of four was reached if the indicator occurred for both civilian participants and both military participants selected for each event. This meant that each indicator could occur a maximum of sixteen times across all of the events. The percentage of each component indicated reflects the number of times all the indicators for a component occurred divided by the maximum amount that indicators could have occurred for that component. The percentage of each concept indicated reflects the number of times all the indicators for a concept occurred divided by the maximum amount that indicators could have occurred for that concept.

# Appendix E: Literature Analysis

**Interorganizational Collaboration in Emergency Management**

A prominent early and influential finding emphasized the need for interorganizational collaboration. Throughout the mid-twentieth century, Fritz, Rayner, and Guskin (1958) observed a general phenomenon during disaster events: an array of resources made up of people, information, and materials flood *into* disaster zones. The authors characterized this convergence as a problem because it made achieving disaster response goals more difficult. They posed two main solutions to the problem: (1) greater control of information acquisition and distribution; and (2) greater control of the disaster site itself. While both suggestions warranted oversight by some type of agency that coordinates people, information, and material to minimize "unnecessary" convergence, the organizations that channelled the people, information, and material had to ensure collaboration among front-line organizations to be successful. While the term "coordination" can suggest anti-collaborative control by a single disaster response coordination agency, such an agency was suggested precisely to fulfil the task of minimizing competition and ensuring collaboration among front-line organizations. Indeed, competition can be understood as the mechanism that explains convergence: A disaster provides an opportunity for organizations to display their capabilities and resources compared to other organizations, thereby enhancing their reputations, which helps bring in future funds. Non-profits, emergency services, and law enforcement agencies flood disaster zones with resources because their focus is not on how to work together to obtain desirable response results, but rather on how salient each organization is vis-à-vis the others. The "coordination agency," explicitly *not* tasked with a front-line function in order to avoid a competitive incentive for itself, is – more than any specific response function – tasked with ensuring collaboration among response organizations.[12]

Lack of interorganizational collaboration as a problem for effective disaster response echoes throughout the subsequent EM literature. In his 2010 overview of disaster research, Drabek (2010) stressed that while it is not uncommon to find sophisticated collaboration mechanisms among branches and/or individuals that exist *within* disaster response organizations, "the thing that hits like a freight train is the marked disorganization among the agencies responding" (148). An introduction to a prominent EM textbook stated that communication among responding organizations is the "Achilles heel" in the field (Haddow, Bullock, and Coppola 2013, 143). An extensive report in *Homeland and Security Affairs* by Donahue and Tuohy (2006) on "lessons never learned" in disaster response stressed the pervasiveness of too little collaboration among organizations. Criticisms of the (mis)management of Hurricane Katrina prominently included the lack of interorganizational collaboration due to an absent "effective" coordination mechanism (Farazmand 2007). In their work on crisis management, Boin, 't Hart, and Kuipers (2007) suggested that contemporary citizens expect their governments to play a role during disaster response (48); based on this assessment, the authors argued that there is a need for a public coordination agency that can induce collaboration among response organizations (50).

While disaster response failures are blamed on a lack of interorganizational collaboration, disaster response "success stories" have been characterized as the result of such collaboration. The largely lauded response to the 2013 Boston Marathon bombing has been studied as a model of intergovernmental collaboration (Hu, Knox, and Kapucu 2014), and Canada's international relief effort immediately after the 2010 Haiti earthquake was praised for the way departments within the federal government worked together (Mamuji 2012). Indeed, the importance of interorganizational collaboration to disaster response is found in an array of studies on EM frameworks conducted by scholars from a variety of disciplines.[13]

EM policy-makers and practitioners have complemented the academic argument for interorganizational collaboration by enacting policies, programs, and organizations that seek to maximize collaboration among disaster response actors. The "professional model" of all-hazards EM that arrived after the "traditional," nuclear attack–focused model – a development discussed at length in chapter 1's EM overview – underscored "the need to integrate activities," where "the police, fire and [Emergency Medical Services] collaborate with the media, the coroner's office, and crisis counselors" (McEntire 2007, 99). The government "coordination agencies" that were developed in Canada and the US throughout the latter half of the twentieth century were mandated to spark interorganizational collaboration and were identified as the mechanisms through which disaster response activities could be integrated. In 1979, the US Federal Emergency Management Agency (FEMA) was created by President Carter (McEntire 2007, 96), and Canadian provinces filled their country's federal void in EM – also discussed in chapter 1 – by establishing their own disaster response coordination organizations throughout the 1970s (Scanlon 1982). Today, FEMA and provincial EM organizations (EMOs) are salient features in their respective governments' bureaucracies, and join a host of other coordination-tasked agencies and individuals at different levels of government.

## Contemporary Criticism of Hierarchy among Disaster Response Organizations

The components of collaboration that emerge from EM are embedded within the contemporary EM literature's critical assessment of hierarchical systems that require clean lines of top-down authority with tightly controlled communication and direction-giving channels. While strict hierarchy may allow a degree of coordination during disaster response when, for example, lead agency X ensures that organization Y does not duplicate organization Z's efforts, the control X exerts over Y and Z's spontaneous actions causes useful disaster response activities to go undone. Allowing Y and Z to work together – to collaborate – as they see fit on the ground, versus merely fulfilling mutually exclusive tasks, enables them to notice in real time previously unidentified tasks that should be prioritized, where another organization's vulnerabilities need to be assuaged, and what other organizations should be directly called in to support or take over a specific task.

The Incident Command System (ICS) is the most commonly used and widely criticized of hierarchical systems that does not allow for such dynamics. The ICS assumes that the fundamental disaster response problems of resource misallocation and miscommunication can be ameliorated through rigorous lines of authority that monopolize specific operational spheres. A formal flow chart,[14] where specific individuals/organizations are responsible for logistics, communications, operations, and so forth, defines the ICS. While the ICS has been shown to work well for its original purpose (mass wildfire response), EM scholars have amassed an impressive set of case study data to demonstrate ICS shortcomings. They show that strict hierarchy can compound and even spark the very misallocation and miscommunication failures that systems like the ICS try to fix, while collegial collaboration and spontaneous organization can ameliorate them (Zurcher 1968; Kendra, Wachtendorf, and Quarantelli 2003; Wachtendorf and Kendra 2005; Voorhees 2008; Groenendaal and Helsloot 2016; Boersma et al. 2014; Jensen and Thompson 2016; Rimstad et al. 2014; Jensen and Waugh 2014). Of special importance to this research project is that these authors do find and/or suggest contexts where the *internal* make-up of an organization merits strict hierarchy – such as the military – while such hierarchy *among* disaster organizations is generally assessed as negative. For example, response effectiveness suffers when an agency mandated with coordinating front-line organizations during disaster response attempts to dictate specific response actions on the front line (i.e., from an EM perspective, coordinate does not – and should note – equal control).[15]

Features that would contrast the strict hierarchy of ICS are (1) decentralized organizations that can act on the ground – within the constraints of the main goal – without first requesting confirmation from a coordination and/or lead agency; (2) horizontal power relationships among representatives from different organizations that do not demand establishing clear superior–subordinate chains of command; and (3) fluid organizational mandates that adapt to context.

## The Use of Purely Empirical New Public Governance

While the management practices of new public governance (NPG) dovetail with the organizational relationships suggested by EM's interorganizational collaboration and civil–military relations' shared responsibility, the motivation within NPG is broader than for the other two approaches. Non-hierarchical public sector management in NPG is rooted in part by a normatively democratic goal for public administration, while the EM and civil–military relations approaches are largely rooted in empirical claims about effectiveness; interorganizational collaboration and shared responsibility in their respective literatures increase effectiveness during disaster response and when militaries work with their civilian counterparts, making effectiveness their raison d'être. NPG, in turn, casts a wide net for potential stakeholders to engage a policy problem and rejects hard superior–subordinate lines of organizational authority, in part to ensure that systematically marginalized voices contribute to the public policy process. Indeed, NPG understands the "policy community" as a phenomenon that transcends government and business to include formal and informal groups from civil society; anyone who has a stake in a policy problem should have a place not only at the policy implementation table, but also at the policy formulation table.[16]

However, as this study's focus is on effectiveness, the parts of NPG that are of interest are those parts that see collaboration in the public sector as not only a normative goal, but as more effective in terms of reaching desirable policy outcomes. Indeed, NPG was identified as an important approach within public administration writ large by this study because the approach stresses its usefulness in facing complex challenges that require input from different levels of government and types of organizations. The normative goal of including diverse viewpoints as an end in itself is less of a concern here than discerning the nature of effective policy implementation.[17] As such, NPG's focus on including civil society and non-profit input for the sake of doing so is not included as what should be looked for when disaster response effectiveness in Canada is assessed. Rather, NPG's prescription for effective policy implementation independent of its normative concerns, and given the main players already established in a policy area, in this case government organizations, is the focus.

# Notes

### Introduction

1 A federal minister mandated with emergency preparedness and emergency management capacity does reside in the federal department of Public Safety Canada. However, both the minister of public safety and the department as a whole face significant operational limitations during times of disaster response that are elaborated upon in chapter 1's overview of emergency management in Canada.
2 Emergency management consists of at least four phases: mitigation, preparedness, response, and recovery.
3 Chapter 1 also engages the debates on the definitions of "disaster" and "emergency management." Practical definitions anchored in legislation are demonstrated to be most useful in this context given the research inquiry's "home base" in public administration.
4 Every results chapter (chapters 3, 4, and 5) compares each disaster response within each chapter's relevant components to provide an overall picture of the presence of, quality of, and barriers to interorganizational collaboration (versus chapters on each case study that focus on one particular event).

### 1. Emergency Management and the Military in Canada

1 The EM presence at the municipal level ranges from powerful agencies in large cities to one component of one individual's job in some rural municipalities/counties (Henstra 2013).
2 While such natural hazards had occurred in the past, the twentieth century saw a rise in both the frequency of natural hazards as well as the public's expectation that government play a role in mitigating them (Kuban 1996; Jones 2007).

3 To be discussed at length in the literature review section of chapter 2.
4 For example, the *Journal of Business Continuity and Emergency Planning*, the Canadian Standards Association's standard on business continuity, and the Manitoba Health's Disaster Management branch, which is both an EMO for the health sector and a business continuity office for the health department.
5 It is common for elected politicians within jurisdictions that carry large EM responsibility to use the opportunity immediately after disasters to propose EM policies that empower their jurisdiction and/or improve their chances of re-election (Johnson, Tunstall, and Penning-Rowsell 2005; Olson, Olson, and Gawronski 1998).
6 This is highly unlikely to occur in the contemporary moment given (1) the many CAF responses to domestic disasters since then, which has regularized appropriate legal channels for deployment, and (2) the 2007 *Emergency Management Act*, which has provided a clearer federal framework for disaster response in general.
7 During the 2010 Hurricane Igor response, Newfoundland and Labrador's Premier Danny Williams walked a fine line between scoring political points by championing his province's resiliency independent of federal support, and losing political points due to almost requesting military support too late to save property. Provincial premiers face a complex incentive matrix during disasters, where expressing provincial strength while requesting CAF support can be contradictory, yet both contain political rewards.
8 Note, however, that EMOs do not enjoy the level of independence and name recognition of arm's-length organizations such as law enforcement agencies or even health authorities. EMOs are generally not situated within a "special operating" framework and are usually ensconced within a government line department (although this may change as provincial EMOs gain more clout and resources given the increasing frequency and impact of many natural hazards).
9 Canada's territories – regional governments in the North without the constitutional clout of the provinces – have largely the same EM capacity as provinces relative to their population size. The RFA process is simpler given that the request is not coming from a "co-sovereign" government, and opportunities for blame avoidance are greater given that responsibility can be passed up to the federal government. However, the regional nature of disasters and the territories' push towards provincial status means that EM dynamics in the territories, from relatively robust EMOs to EM capacity in policy-specific areas (e.g., health departments) largely mirror those of the provinces.
10 The role of CAF as a federal resource and a ubiquitous actor in domestic natural disaster response will be expanded upon at length in the next section.

11 The public safety partner agencies and review bodes that fall under this umbrella are the Canada Border Services Agency (CBSA), the Canadian Security Intelligence Service (CSIS), the Correctional Service of Canada (CSC), the Parole Board of Canada (PBC), the Royal Canadian Mounted Police (RCMP), the RCMP External Review Committee (ERC), the Civilian Review and Complaints Commission for the Royal Canadian Mounted Police (CRCC), and the Office of the Correctional Investigator (OCI).
12 For example, municipal and state EMOs qualify for funds if they establish FEMA backed guidelines and models, such as the Incident Command System.
13 As will be discussed in the results chapters, government officials often characterize provincial and large municipal EOCs as "outward looking," with influence on decisions taken on the ground, while the Government Operations Centre is characterized as merely "feeding the beast," meaning a federal demand for information that may not be useful in terms of making disaster response decisions (Interview 4, 2017).
14 While the 2007 *Emergency Management Act* and a string of twenty-first-century domestic disaster responses have regularized a more formal RFA process, the historical cases of military action without explicit political approval from the federal government are not so rare in Canadian history. Indeed, the now famous Second World War special forces training camp for Allied soldiers in Ontario – mysteriously called "Camp X" – was initiated and operated by the Canadian military with British and American counterparts without the knowledge of the Canadian prime minister, never mind regular members of cabinet (Bryant and Clark 2017).
15 The nature of the RFA process, on paper and in practice, is detailed in the results chapters.
16 In another classic study on the infamous Waco, Texas, tornado, Moore (1958) also observes the link between the intensity of a disaster and the movement of "managing" the disaster up the political authority chain.
17 Semantics are par for the course when it comes to EM legislation and policy; "non-routine emergencies" largely fit the description of "major disasters" in the US *Stafford Act*.
18 While a detailed discourse on disaster definition exists within the operational sphere (see, for example, Boin, 't Hart, and Kuipers 2007), the distinctions between (1) regular, day-to-day emergencies; (2) larger unanticipated emergencies that are the main purview of official EM; and (3) complex emergencies that transcend multiple sectors, are fairly common (although even here different words are often used to label each distinction) (Hale 2013).
19 For an extended discussion on frameworks used to categorize emergency and disaster types, see Handmer and Dovers (2013).

20 While militaries deploy relatively often to support regional governments during disasters, the geographic, frequency, and impact variability of disasters means that any individual regional government rarely works directly with the military. Indeed, such occurrences are rare enough for individual civilian officials and politicians to clearly remember them as distinct moments in their careers.
21 These Department of Natural Defence performance reports are submitted annually and capture all of the domestic disaster responses of the previous year. Reports archived by the Government of Canada since 1997 can be found here: https://publications.gc.ca/site/eng/9.831630/publication.html.
22 The opposite skill, not made particularly salient in humanitarian tasks, is blowing up bridges, a task at which engineers are equally skilled.
23 Combat language is not replaced when describing limitations to potential tasks; "rules of engagement" and – more colloquially – "staying within your arcs" (as in shooting target arcs) are the phrases used.
24 Indeed, the three branches functioned largely as independent organizations until the Government of Canada, under the direction of Prime Minister Pierre Trudeau, fused them into the single Canadian Armed Forces in 1968.
25 Six CF-188 Hornets were deployed on 29 April 2014, followed by a platoon-sized ground force in Central/Eastern Europe, as well as RCN patrols in the Mediterranean Sea; the deployment continued into 2017 (DND and CAF n.d.c).
26 Details on all of these operations can be found on the continuously updated "Current Operations List" on the Government of Canada's National Defence web page (https://www.canada.ca/en/department-national-defence/services/operations/military-operations/current-operations/list.html).
27 While extensive public opinion polls on how Canadians feels about CAF responses to specific domestic events do not exist, media reports on all of the events assessed here include citizen appraisals of CAF's presence in their communities during disasters. Such appraisals are largely positive in each case.

## 2. Assessing Disaster Response through an Original Collaborative Framework

1 Both the US and Australia are geographically large countries that deal with significant natural hazards like those faced by Canada. While the US is a republic with a much larger population, all three countries share the UK as an institutional "parent," federal constitutions, similar economic

and technological developments, and professional militaries that regularly respond to domestic natural disasters.
2. The University of Delaware's Disaster Research Center (DRC) is, along with the University of Colorado Boulder's Natural Hazard Center, one of the primary academic disaster research hubs in the US. The DRC houses one of the most extensive disaster research databases in the world. The database is multidisciplinary (but remains largely within the social sciences), and does not discriminate according to methodology; disciplines and research represented include sociology, public administration, economics, case studies, and longitudinal studies. The database is managed by full-time staff, not limited to American research, and contains physical texts and digital sources (Disaster Research Center, n.d.).
3. Two of these articles are the two previously cited, which Scanlon wrote on domestic disaster response. The third focuses on CAF's contribution to the Hurricane Katrina response, which, as an international mission, keeps the grand total of academic articles that assess the CAF's role in domestic EM to two.
4. Efficiency is, of course, a separate matter.
5. Although not focused on the disaster response context, competition-driven theories play an important role in the literature on understanding the behaviour of public sector organizations (Breton and Wintrobe 2008).
6. Academic EM research developed from the generically titled "disaster research" field as other disciplines joined sociology in studying those events and processes that adversely impact human-created systems.
7. Note that these "measures" or outcomes of effectiveness are multifaceted; they include the development of individual coping mechanisms, the ability of response agencies to learn, as well as the degree to which private and public property is protected. Definitions of effectiveness vary depending on the type of outcome(s) valued. Insurance companies may define effectiveness as minimizing insurable losses; government disaster recovery programs as minimizing non-insurable losses (because government may then have to provide financial aid); local law enforcement agencies based on the number of individuals evacuated; the Salvation Army based on how many people were fed and housed; and so on. Effectiveness as a function of a particular desirable outcome is not the focus here because the disaster response system writ large can be (in)effective in a variety of areas. Rather than identify a specific effectiveness outcome to report that disaster response system is good at X, the goal here is to assess for a broader phenomenon – interorganizational collaboration – that is shown to be a crucial mechanism in achieving desirable outcomes across an array of areas.

8 See Appendix E for an extensive review and analysis of this literature.
9 A strict hierarchy amongst organizations, enforced by a powerful central agency, severely curtails effective disaster response. Appendix E demonstrates the limits of such centralized organizational control.
10 A country's armed forces can include its law enforcement when the police become militarized.
11 See, for example, Herspring 2013; Saideman 2016; Thompson 2010; Simpson 2011; Haley and Lorenzo 2009; Amundson, Lane, and Ferrara 2008; Gaydos and Luz 1994; Tatham, Oloruntoba, and Spens 2012; Gordenker and Weiss 1989; Tatham and Rietjens 2016; Tatham, Spens, and Kovács 2016; Savage et al. 2015.
12 "Strategy" is the formulation of goals and the broad plan to achieve such goals while "tactics" are those actions taken to achieve objectives that add up towards the goal.
13 Two important points regarding shared responsibility should be clarified: (1) Shared responsibility does not suggest changes to the strict hierarchy *within* the military but argues for horizontal collaboration between civilian and military leaders; and (2) shared responsibility does not suggest any changes to the democratic oversight of the military by elected politicians. While shared responsibility observes and prescribes military involvement in the policy formulation process, the ultimate decision-making power rests with the civilian authority.
14 While the importance of organizational role clarity during disaster response is stressed by some (Curnin et al. 2015), the more dominant thread in the literature stresses that an effective response "require[s] much closer relationships, connections, and resources and even a blurring of the boundaries between organizations" (Keast, Brown, and Mandell, 2007, 19).
15 As schools and programs of public policy and/or administration proliferated throughout North America and Europe at the undergraduate and graduate levels, the academic disclaimer of being an "economist by training," or a "political scientist by training" became less common (Botha, Geva-May, and Maslove 2017).
16 An assessment of public policy implementation can, of course, influence policy formulation (and in fact is usually meant to). Indeed, while this research focuses on implementation, the research results will benefit formulation as well as implementation.
17 While the Red Cross, Salvation Army, other non-profits and charities, as well as local companies are all involved in the provision of disaster services, they are almost always funded, overseen, and joined by government. The non-government sectors usually only become main players when individuals and communities enter the disaster recovery phase; in Canada, government runs the formal response (Kuban 1996).

18 See Appendix E for a discussion of the normative elements of NPG and why this book utilizes NPG's non-normative, empirical claims.
19 On the importance of satisfaction, see Bhakta Bhandari, Owen, and Brooks 2014; Schmeltz et al. 2013. On the importance of identifying and dealing with conceptual difference, see Seppänen et al. 2013; Flin and Fruhen 2015; Vidal 2015; Meshkati and Khashe 2015; Gralla, Goentzel, and Fine 2016; Colville, Pye, and Brown 2016; Endsley 2015; Owen et al. 2016; van Laere 2013.
20 The term "bureau" is used in economic theories applied to public sector contexts and is a general term for government organizations.
21 The inter-agency competition that led to the US security agencies' failure to anticipate the 9/11 terrorist attacks is an infamous case of individuals within government organizations not sharing information to maximize outcomes that would enhance their own prestige (National Commission on Terrorist Attacks upon the United States 2004).
22 Economic theories of the bureau, especially the one put forward by Breton and Wintrobe (1986, 2008) include detailed models on how individuals – assumed to be self-interested – within an organization interact with their managers to compound organizational competition (i.e., anti-collaboration). Such details, however, are beyond the scope of this book. The precise internal dynamics that lead to collaboration in the other approaches are not dissected by this research because those approaches do not make assumptions about individual behaviour as the economic theory does, and because this book's focus and context is the effectiveness of public sector organizations in implementing public policy within a policy domain. While this book investigates claims about organizational dynamics, it does not investigate individual psychology itself, and as such does not focus on the claims made about individuals in economic theories of the bureau.
23 See Appendix B for a detailed discussion of the framework, including indicators for each component and the interview rubrics developed to assess for each indicator.
24 The Halifax explosion was so catastrophic that it was studied during the development of the atomic bomb (Scanlon 1988). Prince's influence – albeit often unrealized or uncredited – extends beyond disaster research as well: The theory of social change that he developed in his thesis almost perfectly mirrors later and ostensibly novel theories of policy change that identify "critical junctures," "policy windows," and "focusing events" as key variables in explaining change.
25 Data acquisition, research participants, and the use of interviews are discussed at length in Appendix B.
26 Industrial accidents, such as the infamous 2013 Lac-Mégantic train derailment and explosion, do not make this list because, despite being "non-routine" events, they are often managed by organizations not explicitly

under the EM umbrella, such as Transport Canada. Furthermore, this research is orientated towards the unique dynamic of an armed organization responding to hazardous events not directly caused by humans. Applying this book's results and collaborative framework to non-natural disasters is discussed in the concluding chapter.

27 A discussion on how EM policy and practice has evolved over time can be found in the EM overview section of chapter 1.
28 "Significant" is here defined as the deployment of at least 800 soldiers and/or assets from each of the three CAF branches.
29 Hurricane Igor's most detrimental impact was contained to the island of Newfoundland. The province's full name is used because provincial jurisdiction above and beyond geographic location is important to how policy implementation plays out.
30 Appendix B demonstrates how the selected cases are linked into the collaborative framework.
31 A complete list of sources used for each disaster event is provided in Appendix C.
32 Some forecast models had Igor completely missing Canada.
33 One report noted that "Newfoundlanders appeared dazed by the degree of the damage" (Environment and Climate Change Canada n.d.a). Another report stressed that "Hurricane Igor changed the mindset of the people of the Province; no longer can we look at hurricanes as something that happens elsewhere. The storm will serve as the benchmark against which all future storms will be measured" (Fire and Emergency Services – Newfoundland and Labrador 2011).
34 These numbers were not quite reached, leading to questions about the need for intentional spilling.
35 The 2011 Assiniboine River flood was characterized as a "non-event" for the City of Winnipeg in the province's post-flood Task Force Report.
36 The distinction between evacuation notices and evacuation orders is fuzzy and differs according to each province. The legal clout of government evacuation action and the extent to which coercion was used is examined across the four cases.
37 Significant jurisdictional issues arose regarding which level of government is responsible for what during First Nations' flood management. These issues, although relevant to understanding the nature of Canadian multi-level governance writ large, is beyond the scope of this study.
38 Federal transfers to compensate provincial costs are ongoing. Despite a set funding formula, Ottawa and provincial capitals often debate which specific programs warrant federal reimbursement.
39 This could be because many tasks were not completed close to a human community (i.e., "upriver"). However, every task occurred within some

type of municipal jurisdiction, and likely a rural municipality (RM) if specific communities are not listed.
40 3,000 buildings were flooded and 4,000 businesses were adversely affected.
41 The now infamous removal of firearms from High River homes by the RCMP while residents were evacuated has gained much press. Assessing these actions, however, is beyond the scope of this study.
42 The "whole of government" concept is applied at the federal level during substantive but discrete tasks as well, ranging from humanitarian relief after the 2010 Haiti earthquake to the decade-long combat mission in Afghanistan (Mamuji 2012; Saideman 2016).
43 Unlike the Op LAMA report, this one does not stress "voluntary" evacuation.
44 Overall costs associated with northern wildfires are often less than floods because (1) the forests in which fires occur do not wind through expensive infrastructure as so many rivers do (i.e., development clears forests around urban settings while urban settings develop around rivers); and (2) forests by definition are not highly valuable farms with high crop yields. Response costs separated from damage costs, however, are substantial. The cost of fire suppression programs easily outweighs the other emergency management program costs at the provincial level.

## 3. The *Presence* of Interorganizational Collaboration

1 Since the relevance of specific indicators to each component (e.g., why multi-organizational representation in an Emergency Operations Centre indicates information sharing) are extensively detailed and developed in chapter 2 and Appendix B, justifications for why an indicator is linked to a component will not be replicated in the results chapters. The focus will rather be on the degree to which specific indicators were present and the implications the level of indicators have for answering the research questions.
2 A table detailing the descriptive components and their indicators according to participants from each event is provided at the end of this chapter. A table detailing the descriptive, qualitative, and normative components and their indicators (i.e., capturing all the results) is provided in Appendix D.
3 As is explained in chapter 1's overview of CAF, reservists are CAF members who are not part of the regular force. Reservists are generally civilians that serve part-time (Class A), but they can hold full-time contracts (Class B or C). Most reservists, especially reservist officers, receive similar training to – or, indeed, train with – their regular force counterparts.

4 Unless otherwise indicated, all interviews were conducted by the author between 24 November 2016 and 24 November 2017.
5 While EOCs are theoretically intended only for managing an incident requiring a provincial-level response, they evolved in all provinces studied to be used for "high level meetings" where essential strategies, even during routine times, are discussed. Who is "in" and "out" of the EOC became a useful snapshot of influential players in a province's emergency management system.
6 "O-groups" are update sessions that occur at a set time during a set time interval (i.e., 20:00 every twenty-four hours) and are attended by a unit's leadership to take stock of what has been accomplished since the last O-group, plan tasks until the next O-group, and generally assess how close the operation is to reaching an "end state." The military emphasis on defining an end state, and the implication such an emphasis has on the quality of interorganizational collaboration, will be discussed in chapter 4.
7 Specific examples of each will be discussed during the flexibility and support sections of this chapter. The focus in this section is purely on whether and how information was shared.
8 The implication hazard-specific training requirements had on the quality of interorganizational collaboration is discussed at length in chapter 4's section on integration.
9 An emergency management function includes any function within the phases of preparedness, mitigation, response, and recovery. As most municipalities focused their budgets – which rely on limited revenue generation tools compared to higher levels of government – on services, those that did have fiscal room for emergency management focused on response, such as establishing an EOC.
10 The director of Emergency Management for the City of Brandon has won numerous awards for his emergency management work, including an innovative large-scale exercise that focused on people with disabilities (Neufeld 2012). While all the civilian research participants in this project stated that an organization with CAF's capabilities was required for a response to the disaster event in question, the City of Brandon's approach may warrant further study as the community was one of the hardest hit during an event that saw substantial CAF deployment, and yet barely used CAF resources.
11 The importance of spontaneous, informal, and local networks is discussed at length in chapter 2 and in the section on interorganizational collaboration in the EM literature in Appendix E.
12 Fire and Emergency Services is the EMO for Newfoundland and Labrador.
13 Prime Minister Harper had indicated during his 2006 federal election campaign that Newfoundland and Labrador should be able to keep

non-renewable resource revenue within the province. The prime minister distanced himself from such statements once he had won the election, sparking an anger in the Newfoundland and Labrador premier that – even years after leaving office – appears to be far from subsiding.

14 The RFA process and its implications for interorganizational collaboration will be discussed at length during chapter 4's sections on coordination and integration, as well as in chapter 6.

15 Overall, the process from RFA initiation to RFA confirmation took two days, a process that can often be done in less than one day. One official involved at the federal level on moving RFAs from initiation to confirmation noted that "my fastest was an afternoon" (Interview 4, 2017).

16 CAF's internal response timelines once they have information to deploy are discussed in chapter 4's section on coordination.

17 The 2017 wildland fire response in northern Manitoba, which saw over 6,000 individuals evacuated from First Nations communities, did see questions around the timeliness of RCAF planes. However, this event was not conducive to assessing the impact of informal emergency management links with local reservists as the RFA came straight from the federal Indigenous and Northern Affairs Canada department.

18 Flexibility is built into this section because the one indicator for flexibility – "willingness to reach a decision not originally anticipated by an organization" – dovetailed perfectly with the non-manipulative influence indicator of "capability to change actions of others in a non-manipulative way." Whenever participants were asked questions crafted to probe these "separate" indicators they would repeat themselves, sometimes even using the same examples. A general discussion on flexibility concludes this section.

19 As the CAF response to the Alberta floods were, like every other response in the book, under the legal umbrella of Aid to Civil Authority – as opposed to Aid to Law Enforcement – the CAF officer was not interested in enforcing evacuation orders, but only in checking on the safety of any individuals who had refused the order (Interview 16, 2017).

20 The visual impact of substantial CAF assets such as RCN frigates and RCAF Hercules aircraft on the quality of interorganizational collaboration is assessed in chapter 4's section on trust.

21 Soldiers still receive detailed ROEs as a part of their formal orders before embarking on a specific mission within their domestic operation, but those ROEs are the "direct descendants" of the RFA, which acts as a sort of "matriarch ROE" that constrains CAF action at a broad, overview level. In other words, no ROEs given to individual soldiers by their platoon commanders can allow more leeway in soldier action than what the RFA allows for CAF in general.

22 The role that misunderstandings of the RFA plays in barriers to interorganizational collaboration is discussed in chapter 6's section on conceptual difference. The focus here is on the impact the RFA had as an institutional constraint on the ability of civilian and military organizations to influence each other's actions.
23 This dynamic may transcend CAF and exemplify a universal difference between strategic decision-makers who have to take into account abstract values like "democratic accountability" and on-the-ground responders who are focused on the lives, concerns, and property of people physically in front of them. In the context of public policy, this dynamic also reflects the perennial tension between policy formulation and implementation.
24 Detailed facts and figures on the amount and type of CAF personnel and assets deployed during each event were obtained through archival research and can be found in chapter 2's overview of the events. The results here were obtained through interviewing officials and officers with the focus on how soldiers and civilian emergency managers understood the degree, nature, and effectiveness of the specific types of support provided during the events in which they were involved.
25 Up-to-date physical fitness and medical tests are a basic requirement for any deployment, including domestic ones. There are rare circumstances in which expired tests can be waived, such as when a soldier with expertise essential to a deployment is needed or, in a catastrophic situation were the need for aid trumps the need to verify each soldiers' health. Such waivers, however, require significant bureaucratic footwork within CAF, including sign off by senior commanders above the unit level.
26 While the RCMP have historically not been able to unionize, they do have an association that can advocate on behalf of members to the employer and as such is developing into a quasi-union.
27 The psychological impact on communities of seeing troops on the ground is assessed in chapter 4's section on trust.
28 Mounted infantry platoons usually have four sections of eight soldiers. Each section uses and maintains its own LAV, which often becomes a ninth "section member," complete with personality quirks and affectionate nicknames.
29 There are other transportation modes used by the Army, such as the ubiquitous Medium Support Vehicle System (MSVS), which in civilian-speak are the "green Army trucks" that transport up to thirty soldiers. These trucks, however, are used almost exclusively to simply move troops from point A to point B. As most of their trips can be performed by a civilian equivalent (and are often performed during disasters by locally hired buses, including yellow school buses), they are not viewed by either civilians or military personnel as a CAF contribution to disaster response.

30 Unlike the aforementioned case of Army units being moved closer to Newfoundland prior to RFA approval under the guise of a "troop movement exercise," there is no evidence of RCAF commanders initiating "aircraft movement exercises" in order to move air support closer to disaster sites prior to RFA approval. This may be due to aircraft being considerably more expensive to move – and in much shorter supply – than Army troops. RCAF commanders cannot from a budgetary point of view risk moving aircraft that will not end up being used.
31 While Hercules aircraft are stationed at 17 Wing in Winnipeg, their main function is transport – not evacuation or hazard monitoring – and as such the speed of their arrival is not as paramount as for the other aircraft.
32 Transport Canada aircraft were used to detect and ultimately track down the shooter during the infamous 2014 killings of RCMP officers in Moncton, New Brunswick.
33 A summary of Operation IMPACT and other recent and ongoing CAF operations is provided in chapter 1's CAF overview section.
34 RCAF SAR aircraft work with the civilian Coast Guard and the provinces/municipalities, which are responsible for SAR over sea and land, respectively, to ensure the Canadian SAR system operates 24/7, 365 days a year. The demand on RCAF resources through the SAR system is acutely demonstrated by the fact that any public or private organization, including individual citizens, can call Joint Rescue Coordination Centres (JRCCs) across the country to report a missing plane (or missing people whose last mode of transport was aircraft) and the RCAF, just as any emergency service after a 9/11 call, has to respond. Other CAF branches cannot be directly "called out" by any organization other than the federal government, never mind individual citizens, and as such can much better allocate their resources to match priorities.
35 The role RCN frigates played in effectively reaching coastal communities and adding to the spectrum of disaster relief to include psychosocial support affirms the theory of Scanlon, Stelle, and Hunsberger (2012) that delivering aid from large bodies of water wherever possible is a uniquely effective disaster response as (1) land barriers are avoided and (2) the delivering entity is physically separated from making a foot print on – and thereby taxing or being affected by – the disaster zone.
36 One particularly popular commander during Hurricane Igor eschewed all digital COPs and preferred a white sheet with only the most basic geographical details drawn out and the main military units deployed indicated. There was no evidence after the event that this commander was less successful than commanders with heavily detailed, digital COPs (Interview 5, 2017).
37 CAF orders during Aid to Civil Authority domestic operations do not include a requirement for crypto, but a combination of aging CAF

radios and military habit appears to have allowed crypto to get in the way of smooth and regular military–civilian communications. The crypto infrastructure on CAF radio systems has had similar impacts in other contexts: During an international training mission where the RCN was supposed to provide the communications support, CAF's radio systems were such that the friendly Chilean ships had no idea what the – inexplicably encrypted against their allies – RCN communication entailed. The RCN ultimately had to scramble to find a US ship that had an effective common crypto and relay all information through that ship (Interview 5, 2017).
38 The unit formations here represent an Army chain of command below the Joint Task Force level.
39 Room for bottom-up influence does exist in the apparent hierarchy of the orders format through the "back-briefing" process, which allows each commander receiving orders to back-brief *their* commanders on how the orders were understood, how tasks will be implemented, if any clarifications need to be made, and so on.
40 The details of how much of CAF units' regular operational budgets went to their disaster response versus how much was "subsidized" or recouped later from Public Services and Procurement Canada is an accounting black box that transcends the insight of both this author and his research participants.
41 Conflict between civilian levels of government was also rare, although conflict among civilians *within* the same level of government were noted by a variety of officials. This dynamic will be addressed in chapter 5's section on empire-building.

## 4. The *Quality* of Interorganizational Collaboration

1 As before, disaster responses are compared and contrasted within each component – coordination, integration, satisfaction, and trust – to provide an overall picture of the quality of interorganizational collaboration during domestic disaster response.
2 Avenues for such improvement are discussed in chapter 6.
3 Even routine domestic operations have defined start and end states (ongoing operations have thresholds for end states established) and depend on the continuing existence of a concretely defined and relevant need/hazard/threat that cannot be addressed or ameliorated by a civilian organization.
4 Civilian officials at the provincial and federal levels strayed beyond discussing their specific events and elaborated at length on abstract concepts such as conceptualizing EM as constituting discrete phases versus a fluid

cycle, all-hazards versus hazard-specific response agencies, and the definition of disaster. Military officers did not engage in such theorizing.
5 The Public Safety Canada–DND links on EM diminished even as EM in general at the federal level was being revamped. The general impacts the lack of robust EM capacity within Public Safety Canada had on domestic disaster response is discussed at length in chapter 6.
6 The civilian liaison officer position that linked Public Safety Canada with DND and CAF has been left periodically vacant through the twenty-first century. The impact of distributing this liaison responsibility across individuals on an ad hoc basis, versus having it encapsulated in one specific position (as was General Hillier's intent when he suggested the position during his tenure as chief of defence staff), is discussed in chapter 6.
7 For example, the Royal Canadian Mounted Police or the Canadian Security Intelligence Service. As is discussed in chapter 1, Public Safety Canada does not have an agency specifically mandated with emergency management.
8 While Health Canada and the Public Health Agency of Canada have security-related responsibilities when developing medical countermeasures and/or planning for pandemics, officials and officers with experience at the federal level noted that links between the federal health department and CAF were not robust (Interviews 4, 7, and 19, 2017).
9 This was a delicate balance. Any – even subtle – disapproval of non-evacuees' decisions could have put soldiers in a perceived position of Aid to Law Enforcement (as opposed to Aid to Civil Authority), which in turn would have strayed beyond the RFA and caused political and reputational problems for officials and officers, respectively.
10 The only area where one organization displayed an apparent lack of appreciation for its impact on another organization was when (usually municipal) governments asked for, or found ways to involve, CAF resources beyond the parameters established in the RFA. As discussed in the previous chapter, civilian officials made security-related requests or used CAF resources for routine tasks, both of which transcended CAF's Aid to Civil Authority mandate. However, such actions on the civilian side were conducted in ignorance of the adverse impacts they may have had on CAF, were immediately stopped upon CAF's request, and are therefore understood as a function of the RFA as institutional constraint on collaboration and not seen as evidence for the lack of appreciation by civilian organizations of the impact their activities may have had on CAF.
11 The high levels of joint decision-making on response actions does not contradict the earlier finding of civil–military disagreement on actions that lead to an end state. The former finding indicates the decision-making process and resulting response actions themselves, while the latter finding

indicates how those actions are perceived in the context of the overall disaster response timeline. Civil and military members consistently found ways to make decisions together, but could not collectively categorize the results of those decisions as leading towards an agreed upon point in the future.
12 As was discussed in chapter 3, indicators for information sharing, non-manipulative influence, and flexibility were all impacted in part by the specific hazard of wildfire.
13 As with hazard type, jurisdictional context (i.e., the province where a disaster occurred) had no observable effect.
14 The case of separate civilian and military firefighting units is a good example of the distinction between coordination and integration. Even though the civilian and military firefighting units were directed in a co-ordinated fashion that avoided having units duplicate work or, worse, direct wildfires towards each other, they were not integrated to form units made out of both civilian and military members. Furthermore, there was no evidence to suggest that integrated units would necessarily also have been coordinated. In a phrase, coordination and integration are distinct processes.
15 Drawn out debates on the effectiveness of CAF operations in complex missions such as the Afghanistan conflict – debates that fill textbooks – hinge precisely on how an end state was defined and whether it was reached (Saideman 2016).
16 Soldiers supporting an operation can redeploy without the operation labelled as successful, but an operation cannot be labelled successful without the redeployment of soldiers. In the context of CAF doctrine and liberal democratic norms, the continued presence of an armed force is by itself an indication that an operation – or, indeed, mission – has not yet reached its main goal, which is always to help lay the groundwork for a system that does not require "soldiers in the streets" (Interview 4, 2017).
17 Note that neither military nor civilian organizations removed the reputational component from their measures of effectiveness. The potential success before a response action and assessed success after a response action were influenced by the potential damage the response action itself could cause to reputations, and, as such, "measures of success" were similar in one regard: Neither CAF nor civilian leaders assessed success purely on emergency management outcomes.
18 Many lessons learned *were* incorporated into EMO plans as a matter of best practice (e.g., during the 2013 floods, Alberta employed a successful task force format initially developed during the 2011 Slave Lake fire), but EMOs were not strictly assessed as successful (or unsuccessful) depending on how concretely all lessons learned were applied into their processes, as

was the case for responding military units. Applying lessons learned was a best – albeit not always necessary or feasible – practice in civilian EMOs, while in CAF it was an institutionalized requirement.
19  For a detailed discussion on what a "common situational understanding" entails, see chapter 2's theoretical framework.
20  As discussed in the section on technological support, live tracking of *already deployed* CAF assets did see some discrepancy between the military and civilian sides at the strategic level due to non-interoperable COPs.
21  As noted in the section on integration, military and civilian firefighters largely did not mix within firefighting units, but military and civilian units themselves still needed to communicate with each other. The impact the use of these two separate command systems across different units had on civil–military collaboration during wildfire fighting is assessed in chapter 5's section on conceptual difference.
22  As is discussed at length in chapter 3, the LO function is built to maintain a link with and presence in civilian EOCs during non-response times.
23  The same minister was so impressed by his military counterpart that he sent the senior CAF officer an expensive bottle of scotch after the event (Interview 8, 2017), which – while not captured in this book's methodology – is surely indicative of collegiality.
24  "CADPAT" is short for the distinctly patterned uniforms worn by CAF members.

## 5. The *Barriers* to Interorganizational Collaboration

1  The military–civilian disagreements – assessed in chapter 4 – on what actions should lead to a response's end state occurred in the context of conceptualizing response actions as informing the broader disaster response – versus a specific organization's – outcomes.
2  This dynamic, where CAF commanders were not even considering the potential use of withholding information, is unique to non-combat domestic disaster response. Most other conventional operations would include a degree of information control in CAF's strategy as enemy forces can use information to CAF's detriment. Wildfires, floods, and hurricanes, however, do not exploit information, and civilian response organizations are allies on the "emergency management battlefield."
3  Provincial relations with large municipalities during recent floods in Manitoba are best assessed through Manitoba–Brandon relations, as the City of Winnipeg has been largely shielded from severe flood impacts by the Red River Floodway, which was constructed in 1968 and expanded throughout the 2000s. The Floodway – initially derided as "Duff's Ditch" after Duff

Roblin, the premier who initiated the project – has saved Manitoba an estimated $40 billion in cumulative flood damages since 1968.
4 The provincial–municipal level of openness was not replicated at the provincial–federal or municipal–federal levels, but as gaps in provincial and municipal links to the federal level were due to the civilian federal government's benign incapacity on the emergency management file, such gaps are not assessed as indicators of empire-building and are discussed later in this chapter.
5 The potential reasons for empire-building within the provincial level of government, and the implications of such reasons for the academic literature, are discussed in chapter 6.
6 The GOC is not to be confused with the GoC (the Government of Canada), although officials did – partly tongue-in-cheek – comment that the GOC and the GoC had more in common than their acronyms; they were equally absent during disaster response (Interviews 4 and 17, 2017). Such attitudes indicate that (1) funding, no matter how robust, before or after a disaster event was not valued as highly as putting boots on the ground during the response phase (i.e., despite ongoing calls for such in academic and policy circles, preparedness, mitigation, and recovery are still not seen as equal phases within the emergency management spectrum); and (2) the civilian federal government and CAF were perceived as two distinct entities.
7 For example, in the provincial EMO versus HEM case discussed previously, a HEM member described the EMO's exclusive meetings, but then also described HEM's response of withholding evacuation information, simultaneously implicating HEM in an empire-building indicator.
8 The confirmation of apparent empire-building was obtained indirectly through posing generic questions to EMOs about their meeting attendees and to PHAC and Public Safety Canada about their information flow.
9 This was the case for the junior officers who informed their chain of command – as a matter of routine situational reporting – of municipal leaders prioritizing their own neighbourhoods for hazard reduction efforts.
10 While the interviews did not explicitly probe perceptions of routine emergency response organizations, all officials and officers were asked to paint an overall picture of the disaster response effort. During these initial portions of each interview, participants described an array of organizations and did not shy away from identifying their vulnerabilities. Front-line organizations, however, almost never bore the brunt of such criticisms.
11 See the CAF overview in chapter 1 and the literature discussion in chapter 2 for fuller discussions on CAF culture and doctrine.
12 "Indoctrinated" is not used here in a pejorative sense. "Indoctrination" is used neutrally in CAF to describe the process whereby individuals incorporate a new set of competencies into their skillset and norms into their

Notes to pages 115–20   211

behaviour (processes that occur across all, including civilian, organizations). The association between "indoctrination" and the blind acceptance of ideas, or the inability of individuals to think for themselves, does not dovetail with CAF doctrine.

13  As is elaborated in Appendix B, trust and distrust were treated as two separate components under two separate research queries in order to avoid conceptualizing trust along a simplistic "more versus less" spectrum. The component of trust informed the evaluative assessment, and thereby contributed to indicating the quality of interorganizational collaboration, while the component of distrust informed the normative assessment, and thereby contributed to indicating barriers to interorganizational collaboration. While it makes intuitive sense that as indicators for trust appear, indicators for distrust disappear, the methodology assessed for both independent of each other to ensure the complicated ways in which interorganizational trust may play out were captured.

14  In a similar vein, CAF attempted to change civilian behaviour to only requesting military support sanctioned by the formal RFA, but again, such educating was open about its means, which was education on the RFA, and its ends, which was to allow CAF to achieve joint goals as per the established legal framework.

15  Indeed, an unexpected – and potentially unique – finding of this work is that empire-building can occur in the context of trust (i.e., competition does not require distrust).

16  The sailors themselves, however, were content to continue sharing their rations; another difference between strategic and tactical considerations.

17  The empire-building within civilian levels of government did not result in an unwillingness to leave response actions to others. Even though one organization at times had exclusive or primary access to information or resources, the response abilities of organizations within the same level of government rarely overlapped. For example, HEMs had health system specific information, but did not have the evacuation capacity of EMOs. EMOs, in turn, had access to an array of response organizations through their provincial EOCs, but could implement "pop-up" clinics.

18  The implications of this perspective on the academic support for front-line integration is discussed in chapter 6.

19  Checking in on non-evacuees was immensely stressful for municipal leaders as they often had to play a contradictory role of psychosocial support workers inquiring on the well-being of their residents and of pseudo-compliance officials attempting to enforce provincial evacuation orders. Furthermore, unlike provincial or big city officials, small town leaders were often well acquainted with non-evacuees, inhibiting their ability to act as neutral messengers of state authority. CAF, however, had no

mandate to enforce evacuation orders and simply took over as general assessors of well-being and safety. The role of CAF in checking in on non-evacuees was a largely underreported psychosocial role that was as important to municipalities as the tasks requiring heavy labour.
20 "Arcs" is military-speak for the scope of a mission, operation, trade, or specific task. The word comes from the "left and right of arc," which dictates the area within which soldiers are allowed to fire their weapons.
21 The MND is a civilian-elected official while the CDS is Canada's top soldier.

## 6. Results, Implications, and Recommendations

1 A detailed table of the indicators occurring for each component under each concept can be found in Appendix B.
2 Municipal support can be higher in larger municipalities that have seen prior investments in emergency management (e.g., Calgary).
3 The component of "integration" is not be confused with the word "integrated" used in the main research question. The former is one component out of four under the concept of the quality of interorganizational collaboration, a component that was indicated at low levels, while the latter refers to the overall effectiveness of the CAF–civilian relationship, which was generally high.
4 Most research participants also regarded the partisan presence of the federal government as not important to effective disaster response. While the effect of visiting political leaders on communities in the midst of a disaster is beyond the scope of this study, emergency management officials tended to dismiss prime ministerial visits to the heart of disaster zones as "the dog and pony show" (Interview 2, 2017).
5 The US and Australia are the methodologically preferred comparators given that they are both similarly affluent states with federal systems and established EM programs. Details on the strength of Canada–US–Australia comparison are discussed in chapter 1.
6 CAF's role in response formulation did not contradict the CAF stance of being an *aid* to civil authorities; such a stance was anchored in the RFA as legitimately initiated by civil authority, and any action related to the response – planning or doing – was done under civilian blessing. In other words, Aid to Civil Authority was a concept used to legitimize the military participation in the response, across the strategy-to-tactics spectrum, and not used to constrain military action to the implementation process only.
7 The specifics of military procurement are a function of dynamics in defence politics and industry that transcend the scope of this policy

recommendation. Note, however, that the procurement of new RCAF aircraft is – ironically given that they are the only CAF assets recommended for procurement in this book – especially fraught with delays.

**Appendixes**

1 Collective conflict resolution may seem like it belongs under the "evaluative" section (Research Question 2) of the collaborative framework, but it is kept under the "descriptive" section (Research Question 1) because just as "consultation with stakeholders" in the broader public policy process does not equate to the voices of stakeholders being used in policy formulation, so does the presence of a collective conflict resolution process not mean that conflicts are genuinely a collaborative affair. The indicators of the degree to which the collective conflict resolution process is used can therefore be found in the traits of "satisfaction" and "trust" in the evaluative section.
2 Note that this research is not only hypothesis-testing but also hypothesis-generating, as results may lead to broader hypotheses regarding the nature of military action in federal systems during domestic disaster response (Levy 2008).
3 Note that, as will be seen below, evaluative and normative components were also directly indicated through questions in the interview rubric; the comparison of answers is simply an "extra layer" of assessing the level of these components.
4 "Grouping" was done purely by the researcher when assessing individual responses. Participants did not know other participants and were not interviewed together.
5 A brief history of formal disaster research within academic institutions can be found in the EM literature review section of chapter 2.
6 A group of influential disaster researchers, including Joe Scanlon and Tricia Wachtendorf, conducted interviews with participants during disasters into the twenty-first century. However, such research is largely based within journalism or qualitative sociology, where the limited perspectives and/or crisis-specific emotions and cognitive functions of research participants are not seen as methodological problems but part of what is being assessed. Disaster research within public administration, however, generally interviews participants who can speak about public policies and their implementation independent of their own sense of survival, where the focus is less on the sociological and psychological dynamics of a crisis and more on the organizational and institutional dynamics relevant to public policy implementation during a crisis.
7 This holds true for other descriptive components as well; it may be possible to observe the sharing of emergency plans between some individuals

from some organizations, but not possible to track the full level of sharing in real time throughout an event.

8 The snowball participant recruitment technique, which identifies extra potential participants through the recommendation of existing participants, was particularly useful to this research as some managers and officers intricately involved in managing relevant events were not publicly linked to the event. Recommended participants' confidentiality was protected by not informing recommenders whether individuals they recommended were contacted to participate.

9 All research participant recruitment materials were assessed and cleared by Carleton University's Research Ethics Board.

10 While many military members take on civilian EM roles upon retiring from the military, emergency managers/leaders are either civilian or military during disaster response in Canada; they cannot be both.

11 These participants faced the same interview procedure but were simply asked to speak about each event to identify when one event or response action differed markedly in some respect from the others.

12 The actual success or, indeed, even necessity of coordination agencies in achieving this task has – as discussed in chapter 1's overview of EM – come under recent academic scrutiny. Nevertheless, the potential inability of a coordination agency to achieve interorganizational collaboration does not negate the need for interorganizational collaboration during disaster response.

13 See, for example, Haque 2000; Donahue and Joyce 2001; Newkirk 2001; Henstra and Sancton 2002; Henstra and McBean 2004; Moynihan 2005; Hwacha 2005; Cigler 2005; Wickramasinghe, Bail, and Naguib 2006; Comfort 2007; Foreign Policy Bulletin 2009; French and Raymond 2009; Cigler and Rubin 2009; Somers and Svara 2009; McGuire and Schneck 2010; Henstra 2010; Tatham, Oloruntoba, and Spens 2012; Schwartz et al. 2014; Salmon et al. 2011; Koliba, Mills, and Zia 2011; Kapucu and Demiroz 2011; Parsons 2012; Oh, Okada, and Comfort 2014; Hondula and Krishnamurthy 2014; Arklay 2015.

14 See Appendix A.

15 Note that CAF learned this lesson throughout the twentieth century (in part by observing ineffective British officers during the First and Second World Wars). Senior officers within headquarters or command posts who attempted to assert control over front-line companies and platoons experienced significant strategic blunders, while officers who, after having dictated a lower organization's goal, left it up to that organization to accomplish the goal as it saw fit experienced dramatic successes (Morton 2007).

16 Some scholars – including ones who affirm NPG – have noted that NPG's normative overtones have rendered some its applications blind to issues traditional public administration has been good at, namely accountability (MacDonald and Levasseur 2014).
17 These two outcomes can, of course, occur together, but the part of the NPG approach that is of interest here is its effectiveness claims. Whether NPG's normative goals are achieved or not transcends the scope of this study.

# Bibliography

Agranoff, R., and M. McGuire. 2001. "Big Questions in Public Network Management Research." *Journal of Public Administration Research and Theory* 11 (3): 295–326. https://doi.org/10.1093/oxfordjournals.jpart.a003504.

Agranoff, R., and M. Mcguire. 2003. *Collaborative Public Management: New Strategies for Local Governments*. Washington, DC: Georgetown University Press.

Alison, L., N. Power, C. van den Heuvel, M. Humann, M. Palasinksi, and J. Crego. 2015. "Decision Inertia: Deciding between Least Worst Outcomes in Emergency Responses to Disasters." *Journal of Occupational and Organizational Psychology* 88 (2): 295–321. https://doi.org/10.1111/joop.12108.

Altay, N., and M. Labonte. 2014. "Challenges in Humanitarian Information Management and Exchange: Evidence from Haiti." *Disasters* 38 (S1): 50–72. https://doi.org/10.1111/disa.12052.

Altay, N., and R. Pal. 2014. "Information Diffusion among Agents: Implications for Humanitarian Operations." *Production and Operations Management* 23 (6): 1015–27. http://dx.doi.org/10.1111/poms.12102.

Amiro, B.D., J.B. Todd, B.M. Wotton, K.A. Logan, M.D. Flannigan, B.J. Stoks, J.A. Mason, D.L. Martell, and K.G. Hirsch. 2011. "Direct Carbon Emissions from Canadian Forest Fires, 1959–1999." *Canadian Journal of Forest Research* 31 (3): 512–25. https://doi.org/10.1139/x00-197.

Amundson, E., D. Lane, and D. Ferrara. 2008. "Operation Aftershock: The U.S. Military Disaster Response to the Yogyakarta Earthquake May through June 2006." *Military Medicine* 173 (3): 236–40. https://doi.org/10.7205/milmed.173.3.236.

Andowski, J.L., and A. Auld. 2010. "Hurricane Igor Batters Newfoundland." *Globe and Mail*, 21 September. http://www.theglobeandmail.com/news/national/hurricane-igor-batters-newfoundland/article4326544/.

Andrew, C. 1994. "Federal Urban Activity: Intergovernmental Relations in an Age of Restraint." In F. Frisken, ed., *The Changing Canadian Metropolis: A Public Policy Perspective*. Berkeley: Institute of Governmental Studies Press.

- 1995. "Provincial–Municipal Relations; or Hyper-Fractionalized Quasi-Subordination Revisited." In J. Lightbody, ed., *Canadian Metropolitics: Governing Our Cities*. Toronto: Copp Clark.
Ansell, C., and A. Gash. 2007. "Collaborative Governance in Theory and Practice." *Journal of Public Administration Research and Theory* 18 (4): 543–71. https://doi.org/10.1093/jopart/mum032.
Arklay, T. 2015. "What Happened to Queensland's Disaster Management Arrangements? From 'Global Best Practice' to 'Unsustainable' in 3 Years." *Australian Journal of Public Administration* 74 (2): 187–98. http://dx.doi.org/10.1111/1467-8500.12122.
Arquilla, J., D. Ronfeldt, and M. Zanini. 1999. "Networks, Netwar, and Information-Age Terrorism." In I.O. Lesser, B. Hoffman, J. Arquilla, D. Ronfeldt, M. Zanini, and B.M. Jenkins, eds., *Countering the New Terrorism*. Santa Monica: Rand.
Associated Press. 2010. "Igor's Rains Flood Parts of Northeast Canada." *NBC News*, 21 September. https://www.nbcnews.com/id/wbna39286755.
Attorney General's Department. 2009. *Australian Emergency Management Arrangements*. Canberra: Commonwealth of Australia.
Barton, A.H. 1969. *Communities in Disaster: A Sociological Analysis of Collective Stress Situations*. New York: Doubleday.
Belanger, C. 2000. "Chronology of the October Crisis, 1970, and Its Aftermath." Quebec History. Last modified 23 August 2000. http://faculty.marianopolis.edu/c.belanger/quebechistory/chronos/october.htm.
Bercuson, D.J. 1999. *Blood on the Hills: The Canadian Army in the Korean War*. Toronto: University of Toronto Press.
- 2003. "Canada-US Defence Relations Post-11 September." In D. Carment, F.O. Hampson, and N. Hillmer, eds., *Canada Among Nations 2003: Coping with the American Colossus*. New York: Oxford University Press. https://dspace.ucalgary.ca/bitstream/1880/44573/1/Bercuson_Coping.pdf.
- 2008. *The Fighting Canadians: Our Regimental History from New France to Afghanistan*. Toronto: HarperCollins Canada.
- 2015. *Our Finest Hour*. Toronto: HarperCollins Canada.
Bhakta Bhandari, R., C. Owen, and B. Brooks. 2014. "Organisational Features and Their Effect on the Perceived Performance of Emergency Management Organisations." *Disaster Prevention and Management: An International Journal* 23 (3): 222–42. http://dx.doi.org/10.1108/DPM-06-2013-0101.
Birkland, T.A. 1997. *After Disaster: Agenda Setting, Public Policy, and Focusing Events*. Washington, DC: Georgetown University Press.
Boersma, K., L. Comfort, J. Groenendaal, and J. Wolbers. 2014. "Incident Command Systems: A Dynamic Tension among Goals, Rules and Practice." *Journal of Contingencies and Crisis Management*, 22 (1): 1–4. http://dx.doi.org/10.1111/1468-5973.12042.

Boin, A., P. 't Hart, and S. Kuipers. 2007. "The Crisis Approach." In H. Rodriguez, E. Quarantelli, and R.R. Dynes, eds., *Handbook of Disaster Research*. New York: Springer.

Botha, J. 2015. "Shift Focus from Fighting Forest Fires to Prevention." *Winnipeg Free Press*, 10 July. http://www.winnipegfreepress.com/opinion/analysis/shift-focus-from-fighting-forest-fires-to-prevention-313146491.html.

– 2018. "Services Not Required? Assessing the Need for 'Coordination Agencies' during Disaster Response." In J. Kendra, S.G. Knowles, and T. Wachtendorf, eds., *The New Environmental Crisis: Hazard, Disaster, and the Challenges Ahead*. Newark: University of Delaware Disaster Research Center.

Botha, J., I. Geva-May, and A.M. Maslove. 2017. "Public Policy Studies in North America and Europe." In M. Brans, I. Geva-May, and M. Howlett, eds., *Routledge Handbook of Comparative Policy Analysis*. London: Routledge.

Breton, A. 1998. *Competitive Governments: An Economic Theory of Politics and Public Finance*. New York: Cambridge University Press.

Breton, A., and R. Wintrobe. 1986. "The Bureaucracy of Murder Revisited." *Journal of Political Economy* 94: 905–26. http://www.jstor.org.proxy.library.carleton.ca/stable/1833187.

– 2008. *The Logic of Bureaucratic Conduct: An Economic Analysis of Competition, Exchange, and Efficiency in Private and Public Organizations*. New York: Cambridge University Press.

Bryant, C., and J. Clark. 2017. "What Was Camp X?" *Stuff You Should Know*. Podcast audio 15 June. https://podtail.com/en/podcast/stuff-you-should-know/what-was-camp-x/.

Burton, P.J., M.-A. Parisien, J.A. Hicke, A. Leduc, S. Gauthier, Y. Bergeron, and M. Flannigan. 2008. "Large Fires as Agents of Ecological Diversity in the North American Boreal Forest." *International Journal of Wildland Fire* 17 (6): 754–67. https://doi.org/10.1071/WF07149.

Calgary Herald. 2013. "Graphic: Bow, Elbow and Highwood Flowing Five to 10 Times Normal Rate." *Calgary Herald*, 22 June. http://calgaryherald.com/news/local-news/graphic-bow-elbow-and-highwood-flowing-five-to-10-times-normal-rate.

– 2014a. "The Flood's Tragic Toll: Remembering the Five Lives Lost." *Calgary Herald*, 14 June. https://calgaryherald.com/news/local-news/the-floods-tragic-toll-remembering-the-five-lives-lost.

– 2014b. "High River Flooding Took Toll on Former Mayor." *Calgary Herald*, 23 June. https://calgaryherald.com/news/local-news/high-river-flooding-took-toll-on-former-mayor.

Canadian Army. n.d.a. "Operation LENTUS 13-01: Southern Alberta." Accessed 24 August 2021. https://www.canada.ca/en/department-national-defence/services/operations/military-operations/current-operations/operation-lentus.html.

- n.d.b. "Operation LENTUS 15-02: Northern Saskatchewan." Accessed 24 August 2021. https://www.canada.ca/en/department-national-defence/services/operations/military-operations/current-operations/operation-lentus.html.
- 2013. "Alberta-based Troops Helping Their Neighbours." News Release, 26 June. http://www.army-armee.forces.gc.ca/en/news-publications/western-news-details-secondary-menu.page?doc=alberta-based-troops-helping-their-neighbours/hie8w9kz.

Canadian Interagency Forest Fire Centre. n.d. "Archived Situation Reports." Accessed 24 August 2021. https://www.ciffc.ca/fire-information/archived-situation-reports.

Canadian Joint Operations Command. 2015. *Domestic Operations*.

Canadian Press. 2013a. "Federal Government Announces $200M for 2011 Flood Relief." *CBC News*, 17 September. http://www.cbc.ca/news/canada/manitoba/federal-government-announces-200m-for-2011-flood-relief-1.1857608.

- 2013b. "In Flood Ravaged High River, Canadian Soldiers Find Something Resembling a War Zone." *National Post*, 22 June. https://nationalpost.com/news/canada/in-flood-ravaged-high-river-canadian-soldiers-find-something-resembling-a-war-zone.
- 2015a. "Cost of Saskatchewan Wildfires to Top $100 Million: Wall." *Toronto Sun*, 24 July. http://www.torontosun.com/2015/07/24/cost-of-saskatchewan-wildfires-to-top-100-million-wall.
- 2015b. "Fire, Fumes, and Fear in Western Canada: What You Need to Know." *Globe and Mail*, 6 July. http://www.theglobeandmail.com/news/national/fire-fumes-and-fear-in-western-canada-what-you-need-to-know/article25313346/.
- 2010. "Hurricane Igor Rips in Nfld., Washing Out Roads, Toppling Trees." *Toronto Star*, 22 September. https://www.webcitation.org/5xHOi485p

Carr, J., and J. Jensen. 2015. "Explaining the Pre-Disaster Integration of Community Emergency Response Teams (CERTs)." *Natural Hazards* 77 (3): 1551–71. http://dx.doi.org/10.1007/s11069-015-1664-3.

Catto, N., and S. Tomblin. 2013. "Multilevel Governance Challenges in Newfoundland and Labrador: A Case Study of Emergency Measures." In D. Henstra, ed., *Multilevel Governance and Emergency Management in Canadian Municipalities*. Montreal: McGill-Queen's University Press.

CBC News. 2010a. "Housing Help Needed for Igor Victims: Williams." *CBC News*, 29 September. http://www.cbc.ca/news/canada/newfoundland-labrador/housing-help-needed-for-igor-victims-williams-1.904868.
- 2010b. "Hurricane Igor Attacks Newfoundland." *CBC News*. Last updated 20 September 2020. http://www.cbc.ca/news/canada/newfoundland-labrador/hurricane-igor-attacks-newfoundland-1.935880.

- 2010c. "Igor Outages and Flood Warning: Power Company." *CBC News*, 21 September. http://www.cbc.ca/news/canada/newfoundland-labrador /igor-outages-and-flood-warning-power-company-1.936626.
- 2011a. "Controlled Spill Slows River's Rise." *CBC News*, 15 May. http:// www.cbc.ca/news/canada/controlled-spill-slows-river-s-rise-1.996822.
- 2011b. "Igor Damage Wasn't Fixed Well: Mayor." *CBC News*, 4 October. http://www.cbc.ca/news/canada/newfoundland-labrador/igor -damage-wasn-t-fixed-well-mayor-1.1077052.
- 2011c. "NL's Post-Igor Response Disgusting: Resident." *CBC News*, 16 June. http://www.cbc.ca/news/canada/newfoundland-labrador/n-l-s-post -igor-response-disgusting-resident-1.1022158.
- 2011d. "Water Flows through Manitoba Dike Breach." *CBC News*, 14 May. http://www.cbc.ca/news/canada/manitoba/water-flows-through -manitoba-dike-breach-1.975820.
- 2013. "Special Report: Alberta Flood." *CBC News*, 20 June. http://www.cbc .ca/calgary/features/albertaflood2013/.
- 2015. "Saskatchewan Expects $292M Deficit due to Fires, Oil Price Collapse." *CBC News*, 31 August. http://www.cbc.ca/news/canada/saskatchewan /saskatchewan-expects-292m-deficit-due-to-fires-oil-price-collapse -1.3210239.

CBC News in Review. 2010. "Hurricane Igor Hits Newfoundland." *Curio.ca*, 15 November. https://media.curio.ca/filer_public/a4/9c/a49c1160-1e9c -4a4c-8f3d-00f172fea7ae/nov10igor.pdf.

Charlton, J. 2015. "State of Emergency Declared after 40 People Lose Homes in Montreal Lake." *Saskatoon Star Phoenix*, 4 July. https://thestarphoenix.com /news/local-news/state-of-emergency-declared-after-40-people-lose -homes-in-montreal-lake.

Cigler, B.A. 2005. "The 'Big Questions' of Katrina and the 2005 Great Flood of New Orleans." *Public Administration Review* 67 (s1): 64–76. http://dx.doi .org/10.1111/j.1540-6210.2007.00814.x.

- 2009. "Emergency Management Challenges for the Obama Presidency." *International Journal of Public Administration* 32 (9): 759–66. https://doi .org/10.1080/01900690903016225.

Cigler, B.A., and C.B. Rubin. 2009. "Mainstreaming Emergency Management into Public Administration." *Public Administration Review* 69 (6): 1172–6. http://dx.doi.org/10.1111/j.1540-6210.2009.02074.x.

Cisin, I.H., and W.B. Clark. 1962. "The Methodological Challenge of Disaster Research." In G.W. Baker and D.W. Chapman, eds., *Man and Society in Disaster*. New York: Basic Books.

Colville, I., A. Pye, and A.D. Brown. 2016. "Sensemaking Processes and Weickarious Learning." *Management Learning* 47 (1): 3–13. http://dx.doi. org/10.1177/1350507615616542.

Comfort, L.K. 2007. "Crisis Management in Hindsight: Cognition, Communication, Coordination, and Control." *Public Administration Review* 67: 189–97. https://doi.org/10.1111/j.1540-6210.2007.00827.x.

Constantinides, P. 2013. "The Failure of Foresight in Crisis Management: A Secondary Analysis of the Mari Disaster." *Technological Forecasting and Social Change* 80 (9): 1657–73. https://doi.org/10.1016/j.techfore.2012.10.017.

CTV News. 2010. Igor prompts states of emergency in Newfoundland. *CTV News*, 21 September. http://www.ctvnews.ca/igor-prompts-states-of-emergency-in-newfoundland-1.555232.

– 2011. "Brandon Residents Cope with State of Emergency due to Flooding." *CTV News*, 12 May. http://winnipeg.ctvnews.ca/brandon-residents-cope-with-state-of-emergency-due-to-flooding-1.642314.

Curnin, S., C. Owen, D. Paton, C. Trist, and D. Parsons. 2015. "Role Clarity, Swift Trust and Multi-Agency Coordination." *Journal of Contingencies and Crisis Management* 23 (1): 29–35. https://doi.org/10.1111/1468-5973.12072.

Cutter, S.L., and C. Emrich. 2005. "Are Natural Hazards and Disaster Losses in the U.S. Increasing?" *Eos* 86 (41): 381–9. https://doi.org/10.1029/2005EO410001.

Davison, J. 2013. "Why Alberta's Floods Hit so Hard and Fast." *CBC News*, 22 June. http://www.cbc.ca/news/canada/calgary/why-alberta-s-floods-hit-so-hard-and-fast-1.1328991.

Deep, A. 2015. "Hybrid War: Old Concept, New Techniques." *Small Wars Journal*, 2 March. http://smallwarsjournal.com/jrnl/art/hybrid-war-old-concept-new-techniques.

Department of National Defence (DND). 1997–. *Performance Reports*.

– 2016. "Defence Policy Review Public Consultation Paper." Government of Canada. http://dgpaapp.forces.gc.ca/en/defence-policy-review/consultation-paper.asp.

Department of National Defence and the Canadian Armed Forces (DND and CAF). n.d.a. "Canadian Joint Operations Command." Government of Canada. Accessed 31 August 2021. https://www.canada.ca/en/department-national-defence/corporate/organizational-structure/canadian-joint-operations-command.html.

– n.d.b. "Joint Task Force Atlantic." Government of Canada. Accessed 8 September 2021. http://www.forces.gc.ca/en/operations-regional-jtf-atlantic/jtf-atlantic.page.

– n.d.c. "Operation REASSURANCE." Government of Canada. Accessed 31 August 2021. http://www.forces.gc.ca/en/operations-abroad/nato-ee.page.

– n.d.d. "Regional Joint Task Forces." Government of Canada. Accessed 31 August 2021. http://www.forces.gc.ca/en/operations-regional-jtf/index.page.

- 2011. "The Canadian Forces Complete Flood Mitigation Operation in Manitoba." News Release, 27 May. https://www.canada.ca/en/news/archive/2011/05/canadian-forces-complete-flood-mitigation-operation-manitoba.html.
- 2014. "The Canada-U.S. Defence Relationship." Government of Canada. https://www.canada.ca/en/news/archive/2014/04/canada-defence-relationship.html.
- 2016. *2015–16 Departmental Performance Report*. Government of Canada. http://www.forces.gc.ca/assets/FORCES_Internet/docs/en/about-reports-pubs/2015-2016-dpr-dnd.pdf?t=1.

DiCicco-Bloom, B., and B.F. Crabtree. 2006. "The Qualitative Research Interview." *Medical Education* 40 (4): 314–21. https://doi.org/10.1111/j.1365-2929.2006.02418.x.

Disaster Research Center. n.d.. "The E.L Quarantelli Resource Collection." University of Delaware. Accessed 31 August 2021. https://www.drc.udel.edu/elq-collection/about.

Donahue, A.K., and P.G. Joyce. 2001. "A Framework for Analyzing Emergency Management with an Application to Federal Budgeting." *Public Administration Review* 61 (6): 728–40. http://doi.wiley.com/10.1111/0033-3352.00143.

Donahue, A.K., and R.V. Tuohy. 2006. "Lessons Never Learned: A Study of the Lessons of Disasters, Why We Repeat Them and How We Can Learn from Them." *Homeland and Security Affairs* 2, Article 4. https://www.hsaj.org/articles/167.

Drabek, T.E. 1986. *Human System Responses to Disaster: An Inventory of Sociological Findings*. New York: Springer.
- 2010. *The Human Side of Disaster*. Boca Raton, FL: CRC Press.

Emerson, K., and T. Nabatchi. 2015. *Collaborative Governance Regimes*. Washington, DC: Georgetown University Press.

Endsley, M.R. 2015. "Situation Awareness Misconceptions and Misunderstandings." *Journal of Cognitive Engineering and Decision Making* 9 (1): 4–32. http://dx.doi.org/10.1177/1555343415572631.

Environment and Climate Change Canada. n.d.a. "Canada's Top Ten Weather Stories for 2010: Story Two – Vigorous Igor." Accessed 31 August 2021. https://www.ec.gc.ca/meteo-weather/default.asp?lang=En&n=BDE98E0F-1.
- n.d.b. "Canada's Top Ten Weather Stories for 2011: Story One – Historical Flood Fights in the West." Accessed 31 August 2021. https://www.ec.gc.ca/meteo-weather/default.asp?lang=En&n=0397DE72-1.
- n.d.c. "Canada's Top Ten Weather Stories for 2013: Alberta's Flood of Floods." Accessed 31 August 2021. https://www.ec.gc.ca/meteo-weather/default.asp?lang=En&n=5BA5EAFC-1&offset=2&toc=show.

Eyre, W. 2015. "A 'Thank You' from Your JTFW Commander." Government of Canada, 1 September. http://www.army-armee.forces.gc.ca/en/news-publications/western-news-details-secondary-menu.page?doc=a-thank-you-from-your-jtfw-commander/idw3erqh [Link accessed 2018].

Farazmand, A. 2007. "Learning from the Katrina Crisis: A Global and International Perspective with Implications for Future Crisis Management." *Public Administration Review* 67 (s1): 149–59. https://doi.org/10.1111/j.1540-6210.2007.00824.x.

Federal Emergency Management Agency (FEMA). 2018. "Continuity Guidance Circular." FEMA National Continuity Programs. https://www.fema.gov/sites/default/files/2020-10/continuity-guidance-circular-2018.pdf.

Fire and Emergency Services – Newfoundland and Labrador. 2011. *2010–2011 Annual Report*. House of Assembly Newfoundland and Labrador. http://www.assembly.nl.ca/business/electronicdocuments/FireEmergencyServices-AnnualReport2010-2011.pdf.

Flin, R., and L. Fruhen. 2015. "Managing Safety: Ambiguous Information and Chronic Unease." *Journal of Contingencies and Crisis Management* 23 (2): 84–9. https://doi.org/10.1111/1468-5973.12077.

Foreign Policy Bulletin. 2009. "Emergency Management Agreements with Canada and Mexico Updated." *Foreign Policy Bulletin* 19 (1): 256–68. https://doi.org/10.1017/S1052703609000604.

Form, W., and S. Nosow. 1958. *Community in Disaster*. New York: Harper.

Frederickson, H.G. 1991. "Toward a Theory of the Public for Public Administration." *Administration & Society* 22 (4): 395–417. https://doi.org/10.1177/009539979102200401.

– 2012. *The Public Administration Theory Primer*. Boulder, CO: Westview Press.

Freedman, L. 2013. *Strategy: A History*. New York: Oxford University Press.

French, P.E., and E.S. Raymond. 2009. "Pandemic Influenza Planning: An Extraordinary Ethical Dilemma for Local Government Officials." *Public Administration Review* 69 (5): 823–30. http://dx.doi.org/10.1111/j.1540-6210.2009.02032.x.

Fritz, C., J.F. Rayner, and S.L. Guskin. 1958. *Behaviour in an Emergency Shelter: A Field Study of 800 Persons Stranded in a Highway Restaurant During a Heavy Snow Storm*. Washington, DC: National Academy of Sciences, National Research Group.

Garson, J., and Canadian Press. 2013. "Heavy Rains in Southern Alberta Force Mandatory Evacuations in Areas of Calgary and Surroundings." *National Post*, 20 June. https://nationalpost.com/news/canada/floods-wash-away-roads-in-canmore-close-trans-canada-highway-and-trigger-emergencies-across-southern-alberta.

Gaydos, J.C., and G.A. Luz. 1994. "Military Participation in Emergency Humanitarian Assistance." *Disasters* 18 (1): 48–57. https://doi.org/10.1111/j.1467-7717.1994.tb00284.x.

Geddes, B. 2003. "How the Cases You Choose Affect the Answers You Get: Selection Bias and Related Issues." In *Paradigms and Sand Castles: Theory Building and Research Design in Comparative Politics*. Ann Arbor: University of Michigan Press.

Gerson, J. 2013. "Calgary Residents Grapple with 'Surreal' Devastation as Albertans Lose Cars, Homes in Massive Floods." *National Post*, 21 June. https://nationalpost.com/news/canada/calgary-residents-grapple-with-surreal-devastation-as-albertans-lose-cars-homes-in-massive-floods.

Giles, D. 2015. "Increased Wildfire Behaviour Expected in Saskatchewan due to Weather." *Global News*, 9 July. http://globalnews.ca/news/2100771/some-progress-made-as-soldiers-help-fight-saskatchewan-wildfires/.

Global Affairs Canada. 2017. "Address by Minister Freeland on Canada's Foreign Policy Priorities." Speech, Ottawa, 6 June. Government of Canada. https://www.canada.ca/en/global-affairs/news/2017/06/address_by_ministerfreelandoncanadasforeignpolicypriorities.html.

Goodale, R. 2016. *Public Safety Canada 2016–17: Report on Plans and Priorities*. Ottawa: Her Majesty the Queen in Right of Canada. https://www.publicsafety.gc.ca/cnt/rsrcs/pblctns/rprt-plns-prrts-2016-17/index-en.aspx.

Gordenker, L., and T.G. Weiss. 1989. "Humanitarian Emergencies and Military Help: Some Conceptual Observations." *Disasters* 13 (2): 118–33. https://doi.org/10.1111/j.1467-7717.1989.tb00703.x.

Government of Alberta. n.d. "Alberta 2013 Floods Review." http://www.alberta.ca/flood-2013.cfm [Link accessed 2018].

– 2013a. "Alberta Flood Mapping." Last updated 21 October 2020. https://open.alberta.ca/opendata/gda-2ae32b0d-c6f9-4e1b-81ab-6fdecc728e28.

– 2013b. "Update 5: Government Continues to Respond to Flooding Emergency." Government News, 23 June. http://www.alberta.ca/release.cfm?xID=3439971DE6A21-FD13-B8D7-01FE9183DB16ACAC.

– 2014. "Updated Provincial Flood Statistics." Government News, 6 June. http://www.alberta.ca/release.cfm?xID=3659373BD1D92-CFDB-0D72-DB4C37C4D35EBCFE.

Government of Canada. n.d. "Population of the Federal Public Service." Treasury Board of Canada Secretariat. Accessed 1 September 2021. https://www.canada.ca/en/treasury-board-secretariat/services/innovation/human-resources-statistics/population-federal-public-service.html.

– 2010. "The Canadian Forces Complete Relief Operation in Newfoundland and Labrador." News Release, 6 October. https://www.canada.ca/en/news/archive/2010/10/canadian-forces-complete-relief-operation-newfoundland-labrador.html.

– 2015. "Emergency Management Organizations." Get Prepared. Last updated 15 January 2015. https://www.getprepared.gc.ca/cnt/rsrcs/mrgnc-mgmt-rgnztns-en.aspx.

Government of Manitoba. n.d.a "Manitoba Flood Facts." Flood Information. Accessed 1 September 2021. http://www.gov.mb.ca/flooding/history/index.html.
- n.d.b "Portage Diversion." Manitoba Infrastructure. Accessed 1 September 2021. https://www.gov.mb.ca/mit/wms/pd/index.html.
- 2011a. "Flood Bulletin #40." News Releases, 9 May. https://news.gov.mb.ca/news/print,index.html?item=11448.
- 2011b. "Flood Bulletin #41." News Releases, 10 May. https://news.gov.mb.ca/news/index.html?item=11454&posted=2011-05-10.
- 2011c. "Flood Bulletin #43." News Releases, 11 May. https://news.gov.mb.ca/news/index.html?archive=month&item=11461.
- 2013. *2011 Flood: Technical Review of Lake Manitoba, Lake St. Martin and Assiniboine River Water Levels*. Manitoba Infrastructure. http://www.gov.mb.ca/mit/floodinfo/floodproofing/reports/pdf/assiniboine_lakemb_lsm_report_nov2013.pdf.
Government of Saskatchewan. n.d. "Emergency Management Planning for Communities." Accessed 2 September 2021. http://www.saskatchewan.ca/residents/environment-public-health-and-safety/emergency management.
Gralla, E., J. Goentzel, and C. Fine. 2016. "Problem Formulation and Solution Mechanisms: A Behavioral Study of Humanitarian Transportation Planning." *Production and Operations Management* 25 (1): 22–35. http://dx.doi.org/10.1111/poms.12496.
Granatstein, J.L. 1990. *Canada's War: The Politics of the MacKenzie King Government, 1939–1945*. Toronto: University of Toronto Press.
- 2002. *Canada's Army: Waging War and Keeping the Peace*. Toronto: University of Toronto Press.
- 2005. *The Generals: The Canadian Army's Senior Commanders in the Second World War*. Calgary: University of Calgary Press.
- 2014. *The Greatest Victory: Canada's One Hundred Days, 1918*. Oxford: Oxford University Press.
- 2015. *The Best Little Army in the World: The Canadians in Northwest Europe, 1944–1945*. Toronto: HarperCollins Canada.
- 2016. *The Weight of Command: Voices of Canada's Second World War Generals and Those Who Knew Them*. Vancouver: UBC Press.
Grieve, M., and L. Turnbull. 2013. "Emergency Planning in Nova Scotia." In D. Henstra, ed., *Multilevel Governance and Emergency Management in Canadian Municipalities*. Montreal: McGill-Queen's University Press.
Groenendaal, J., and I. Helsloot. 2016. "A Preliminary Examination of Command and Control by Incident Commanders of Dutch Fire Services during Real Incidents." *Journal of Contingencies and Crisis Management* 24 (1): 2–13. https://doi.org/10.1111/1468-5973.12096.

Gustafson, P.E. 1998. "Gender Differences in Risk Perception: Theoretical and Methodological Perspectives." *Risk Analysis* 18 (6): 805–11. https://doi.org/10.1111/j.1539-6924.1998.tb01123.x.

Gutro, R. 2010. "Hurricane Season 2010: Tropical Storm Igor (Atlantic Ocean)." NASA, 24 September. http://www.nasa.gov/mission_pages/hurricanes/archives/2010/h2010_Igor.html.

Haddow, G., J. Bullock, and D.P. Coppola, eds. 2013. *Introduction to Emergency Management*, 5th ed. Oxford: Butterworth-Heinemann.

Hale, G. 2013. "Emergency Management in Albera: A Study in Multilevel Governance." In D. Henstra, ed., *Multilevel Governance and Emergency Management in Canadian Municipalities*. Montreal: McGill-Queen's University Press.

Haley, F.T., and R.A. DeLorenzo. 2009. "Military Medical Assistance Following Natural Disasters: Refining the Rapid Response." *Prehospital and Disaster Medicine* 24 (1): 9–10. https://doi.org/10.1017/S1049023X00006476.

Handmer, J., and S. Dovers. 2013. *Handbook of Disaster Policies and Institutions*. New York: Routledge.

Haque, C.E. 2000. "Risk Assessment, Emergency Preparedness and Response to Hazards: The Case of the 1997 Red River Valley Flood, Canada." *Natural Hazards* 21 (2): 225–45. http://dx.doi.org/10.1023/A:1008108208545.

Harrald, J.R. 2012. "Intended and Unintended Outcomes of Actions Taken since 9/11." In C.B. Rubin, ed., *Emergency Management: The American Experience 1900–2010*, 2nd ed. Boca Raton, FL: CRC Press.

Hartfiel, R.M. 2010. "Planning without Guidance: Canadian Defence Policy and Planning, 1993–2004." *Canadian Public Administration* 53 (3): 323–49. https://doi.org/10.1111/j.1754-7121.2010.00139.x.

Henstra, D. 2010. "Evaluating Local Government Emergency Management Programs: What Framework Should Public Managers Adopt?" *Public Administration Review* 70 (2): 236–46. https://doi.org/10.1111/j.1540-6210.2010.02130.x.

– 2011. "The Dynamics of Policy Change: A Longitudinal Analysis of Emergency Management in Ontario, 1950–2010." *Journal of Policy History* 23 (3): 399–428. https://doi.org/10.1017/S0898030611000169.

– ed. 2013. *Multilevel Governance and Emergency Management in Canadian Municipalities*. Montreal: McGill-Queen's University Press.

Henstra, D., and G. McBean. 2004. *The Role of Government in Services for Natural Disaster Mitigation*. Institute for Catastrophic Loss Reduction. http://www.iclr.org/images/The_Role_of_Government_in_Services_for_Natural_Disaster_Mitigation.pdf.

Henstra, D., and A. Sancton. 2002. *Mitigating Catastrophic Losses: Policies and Policy-Making at Three Levels of Government in the United States and Canada.*

Institute for Catastrophic Loss Reduction. http://www.iclr.org/images/Mitigating_catastrophic_losses.pdf.

Herrero-Fernández, D., P. Macía-Guerrero, L. Silvano-Chaparro, L. Merino, and E.C. Jenchura. 2016. "Risky Behavior in Young Adult Pedestrians: Personality Determinants, Correlates with Risk Perception, and Gender Differences." *Transportation Research Part F: Psychology and Behaviour* 36 (1): 14–24. http://dx.doi.org/10.1016/j.trf.2015.11.007.

Herspring, D.R. 2013. *Civil–Military Relations and Shared Responsibility*. Baltimore: Johns Hopkins University Press.

Hodgetts, J.E. 1997. "The Intellectual Odyssey of Public Administration in English Canada." *Canadian Public Administration* 40 (2): 171–85. https://doi.org/10.1111/j.1754-7121.1997.tb01504.x.

Hoffman, F.G. 2007. *Conflict in the 21st Century: The Rise of Hybrid Wars*. Potomac Institute for Policy Studies. http://www.potomacinstitute.org/images/stories/publications/potomac_hybridwar_0108.pdf.

Hondula, D., and R. Krishnamurthy. 2014. "Emergency Management in the Era of Social Media." *Public Administration Review* 74 (2): 274–7. https://doi.org/10.1111/puar.12184.

Hu, Q., C. Knox, and N. Kapucu. 2014. "What Have We Learned since September 11, 2001? A Network Study of the Boston Marathon Bombings Response." *Public Administration Review* 74 (6): 698–712. https://doi.org/10.1111/puar.12284.

Huxham, C., S. Vangen, C. Huxham, and C. Eden. 2006. "The Challenge of Collaborative Governance." *Public Management: An International Journal of Research and Theory* 2 (3): 337–58. https://doi.org/10.1080/14719030000000021.

Hwacha, V. 2005. "Canada's Experience in Developing a National Disaster Mitigation Strategy: A Deliberative Dialogue Approach." *Mitigation and Adaptation Strategies for Global Change* 10 (3): 507–23. https://doi.org/10.1007/s11027-005-0058-3.

Ibrahim, D. 2014. "A Look Back at Devastating Hurricane Igor." *The Weather Network*, 21 September. http://www.theweathernetwork.com/news/articles/a-look-back-at-devastating-hurricane-igor/13153.

Iype, M., and B. Bouzane. 2010. "Newfoundland Cleans Up after Hurricane Igor." *National Post*, 21 September. https://nationalpost.com/news/newfoundland-cleans-up-after-hurricane-igor.

Jensen, J., and S. Thompson. 2016. "The Incident Command System: A Literature Review." *Disasters* 40 (1): 158–82. https://doi.org/10.1111/disa.12135.

Jensen, J., and W.L. Waugh. 2014. "The United States' Experience with the Incident Command System: What We Think We Know and What We Need

to Know More About." *Journal of Contingencies and Crisis Management* 22 (1): 5–17. https://doi.org/10.1111/1468-5973.12034.
Jiwani, F.N., and T. Krawchenko. 2014. "Public Policy, Access to Government, and Qualitative Research Practices: Conducting Research within a Culture of Information Control." *Canadian Public Policy* 40 (1): 57–66. http://dx.doi.org/10.3138/cpp.2012-051.
Johnson, C.L., S.M. Tunstall, and E.C. Penning-Rowsell. 2005. "Floods as Catalysts for Policy Change: Historical Lessons from England and Wales." *International Journal of Water Resources Development* 21 (4): 561–75. https://doi.org/10.1080/07900620500258133.
Johnson, E.A., K. Miyanishi, and S. Bridge. 2001. "Wildfire Regime in the Boreal Forest and the Idea of Suppression and Fuel Build Up." *Conservation Biology* 5 (6): 1554–7. https://doi.org/10.1046/j.1523-1739.2001.01005.x.
Johnston, D.L. 2016. *The Idea of Canada: Letters to a Nation*. Toronto: Signal.
Jones, D.T. 2010. "Canada and the United States: Civil Military Relations." *American Diplomacy*. https://americandiplomacy.web.unc.edu/2010/06/canada-and-the-united-states-civil-military-relations/.
Jones, R. 2007. "A Nation Comes Together: A Few Pivotal Events Changed the Way Australia Responds to Its Crises." In J. Keeney and K. Lobley, eds., *In Case of Emergency: How Australia Deals with Disasters and the People Who Confront the Unexpected*. Artarmon: Design Masters Press.
Juillet, L., and Koji, J. 2013. "Policy Change and Constitutional Order: Municipalities, Intergovernmental Relations, and the Recent Evolution of Canadian Emergency Management Policy." In D. Henstra, ed., *Multilevel Governance and Emergency Management in Canadian Municipalities*. Montreal: McGill-Queen's University Press.
Kapucu, N. 2009a. "Interorganizational Coordination in Complex Environments of Disasters: The Evolution of Intergovernmental Disaster Response Systems." *Journal of Homeland Security and Emergency Management* 6 (1). http://www.doi.org/10.2202/1547-7355.1498.
– 2009b. "Leadership under Stress: Presidential Roles in Emergency and Crisis Management in the United States." *International Journal of Public Administration* 32 (9): 767–72. https://doi.org/10.1080/01900690903017025.
Kapucu, N., and F. Demiroz. 2011. "Measuring Performance for Collaborative Public Management Using Network Analysis Methods and Tools." *Public Performance & Management Review* 34 (4): 549–79. http://www.doi.org/10.2753/PMR1530-9576340406.
Kapucu, N., and V. Garayev. 2013. "Designing, Managing, and Sustaining Functionally Collaborative Emergency Management Networks." *The American Review of Public Administration* 43 (3): 312–30. http://dx.doi.org/10.1177/0275074012444719.

- 2014. "Structure and Network Performance: Horizontal and Vertical Networks in Emergency Management." *Administration & Society* 48 (8): 931–61. https://doi.org/10.1177/0095399714541270.
- Kapucu, N., and Q. Hu. 2016. "Understanding Multiplexity of Collaborative Emergency Management Networks." *The American Review of Public Administration* 46 (4): 399–417. https://doi.org/10.1177/0275074014555645.
- Kasurak, P.C. 1982. "Civilianization and the Military Ethos: Civil-Military Relations in Canada." *Canadian Public Administration* 25 (1): 108–29. https://doi.org/10.1111/j.1754-7121.1982.tb02067.x.
- Keast, R., K. Brown, and M. Mandell. 2007. "Getting the Right Mix: Unpacking Integration Meanings and Strategies." *International Public Management Journal* 10 (1): 9–33. https://doi.org/10.1080/10967490601185716.
- Kendra, J., T. Wachtendorf, and E. Quarantelli. 2003. "The Evacuation of Lower Manhattan by Water Transport on September 11: An Unplanned 'Success.'" *Joint Commission of Quality and Safety* 29 (6): 316–18. https://doi.org/10.1016/S1549-3741(03)29036-5.
- Killian, L.M. 1956. *An Introduction to Methodological Problems of Field Studies in Disasters: A Special Report*. Washington, DC: National Research Council.
- Kirschenbaum, A. 2004. "Measuring the Effectiveness of Disaster Management Organizations." *International Journal of Mass Emergencies and Disasters* 22 (1): 75–102. www.ijmed.org/articles/389/download/.
- Knowles, S.G. 2011. *The Disaster Experts Mastering Risk in Modern America*. Philadelphia: University of Philadelphia Press.
- Koliba, C., R. Mills, and A. Zia. 2011. "Accountability in Governance Networks: An Assessment of Public, Private, and Nonprofit Emergency Management Practices Following Hurricane Katrina." *Public Administration Review* 71 (2): 210–20. https://doi.org/10.1111/j.1540-6210.2011.02332.x.
- Kuban, R. 1996. "The Role of Government in Emergency Preparedness." *Canadian Public Administration* 39 (2): 239–44. https://doi.org/10.1111/j.1754-7121.1996.tb00130.x.
- La Ronge EMS (@LaRongeEMS). 2015. "State of emergency in place once again 4 Sikichew & Clam Lake Bridge. Immediate evacuation order in place. Forest fires threatening again." Twitter, 24 June, 8:08 p.m. https://twitter.com/larongeems/status/613861371457875968.
- Leach, W.D. 2006. "Collaborative Public Management and Democracy: Evidence from Western Watershed Partnerships." *Public Administration Review* 66 (s1): 100–10. https://doi.org/10.1111/j.1540-6210.2006.00670.x.
- Lee, J. 2011. "Group Value and Intention to Use – A Study of Multi-Agency Disaster Management Information Systems for Public Safety." *Decision Support Systems* 50 (2): 404–14. https://doi.org/10.1016/j.dss.2010.10.002.
- Leikas, S., M. Lindeman, K. Roininen, and L. Lähteenmäki. 2007. "Food Risk Perceptions, Gender, and Individual Differences in Avoidance and

Approach Motivation, Intuitive and Analytic Thinking Styles, and Anxiety." *Appetite* 48 (2): 232–40. http://dx.doi.org/10.1016/j.appet.2006.09.009.

Levy, J. 2008. "Case Studies: Types, Designs, and Logics of Inference." *Conflict Management and Peace Science* 25 (1): 1–18. http://dx.doi.org/10.1080/07388940701860318.

Lindsay, J. 2014. "The Power to React: Review and Discussion of Canada's Emergency Measures Legislation." *The International Journal of Human Rights* 18 (2): 159–77. https://doi.org/10.1080/13642987.2014.889392.

MacDonald, F., and K. Levasseur. 2014. "Accountability Insights from the Devolution of Indigenous Child Welfare in Manitoba." *Canadian Public Administration* 57 (1): 97–117. https://doi.org/10.1111/capa.12052.

Malcolmson, P., and R. Myers. 2012. *The Canadian Regime*, 5th ed. Toronto: University of Toronto Press.

Mamuji, A. 2012. "Canadian Military Involvement in Humanitarian Assistance: Progress and Prudence in Natural Disaster Response." *Canadian Foreign Policy Journal* 18 (2): 208–24. https://doi.org/10.1080/11926422.2012.709054.

Mandell, M., and T. Steelman. 2003. "Understanding What Can Be Accomplished through Interorganizational Innovations: The Importance of Typologies, Context and Management Strategies." *Public Management Review* 5 (2): 197–224. https://doi.org/10.1080/1461667032000066417.

McEntire, D.A. 2007. *Disaster Response and Recovery: Strategies and Tactics for Resilience.* Hoboken, NJ: Wiley.

McEntire, D.A., and J. Lindsay. 2012. "One Neighborhood, Two Families: A Comparison of Intergovernmental Emergency Management." *Journal of Emergency Management* 10 (2): 93–107. http://dx.doi.org/10.5055/jem.2012.0090.

McGuire, M. 2006. "Collaborative Public Management: Assessing What We Know and How We Know It." *Public Administration Review* 66 (s1): 33–43. https://doi.org/10.1111/j.1540-6210.2006.00664.x.

McGuire, M., and D. Schneck. 2010. "What If Hurricane Katrina Hit in 2020? The Need for Strategic Management of Disasters." *Public Administration Review* 70 (s1): 201–7. https://doi.org/10.1111/j.1540-6210.2010.02273.x.

McLuckie, B.F. 1977. *Italy, Japan and the United States: Effects of Centralization on Disaster Responses 1964–1969.* Columbus: Disaster Research Center, Ohio State University.

Meshkati, N., and Y. Khashe. 2015. "Operators' Improvisation in Complex Technological Systems: Successfully Tackling Ambiguity, Enhancing Resiliency and the Last Resort to Averting Disaster." *Journal of Contingencies and Crisis Management* 23 (2): 90–6. https://doi.org/10.1111/1468-5973.12078.

Miao, X., D. Banister, and Y. Tang. 2013. "Embedding Resilience in Emergency Resource Management to Cope with Natural Hazards." *Natural Hazards* 69 (3): 1389–404. http://dx.doi.org/10.1007/s11069-013-0753-4.

Miller, D.Y., and R.W. Cox. 2014. *Governing the Metropolitan Region: America's New Frontier*. New York: Routledge.

Moore, H.E. 1958. *Tornadoes over Texas: A Study of Waco and San Angelo in Disaster and Its Impact upon the Study of Disaster*. Austin: University of Texas Press.

Morioka, Rika. 2014. "Gender Difference in the Health Risk Perception of Radiation from Fukushima in Japan: The Role of Hegemonic Masculinity." *Social Science & Medicine* 107: 105–12. http://dx.doi.org/10.1016/j.socscimed.2014.02.014.

Morris, J.C., E.D. Morris, and D.M. Jones. 2007. "Reaching for the Philosopher's Stone: Contingent Coordination and the Military's Response to Hurricane Katrina." *Public Administration Review* 67 (s1): 94–106. http://dx.doi.org/10.1111/j.1540-6210.2007.00818.x.

Mortillaro, N. 2015. "Incredible Images of Fires Raging across Western Canada." *Global News*, 6 July. http://globalnews.ca/news/2094173/in-photos-fires-rage-across-western-canada-cities-blanketed-in-smoke/.

Morton, D. 2003. *Understanding Canadian Defence*. Toronto: Penguin Canada.

– 2007. *A Military History of Canada*, 5th ed. Toronto: McClelland & Stewart.

Moynihan, D.P. 2005. "Homeland Security and the U.S. Public Management Policy Agenda. *Governance* 18 (2): 171–96. https://doi.org/10.1111/j.1468-0491.2005.00272.x.

National Commission on Terrorist Attacks upon the United States. 2004. *The 9/11 Commission Report: Final Report of the National Commission on Terrorist Attacks upon the United States*. New York: Norton.

National Oceanic and Atmospheric Administration. 2010. *Hurricane Igor: September 20–21, 2010*. Weather Prediction Center. http://www.wpc.ncep.noaa.gov/tropical/rain/igor2010.html.

Natural Resources Canada. n.d.a. *Canadian Wildland Fire Information System*. Accessed 7 September 2021. https://cwfis.cfs.nrcan.gc.ca/.

– n.d.b. *National Wildland Fire Situation Report*. Accessed 7 September 2021. https://cwfis.cfs.nrcan.gc.ca/report.

Neufeld, C. 2012. "Brandon Emergency Team Gets Award." *Discover Westman*, 4 August. https://www.discoverwestman.com/local/brandon-emergency-team-gets-award.

Newell, D. 2010. "Military to the Rescue in Newfoundland." *Winnipeg Free Press*, 25 September. https://www.winnipegfreepress.com/canada/military-to-the-rescue-in-newfoundland-103776999.html.

Newfoundland and Labrador Statistics Agency. 2010. *Hurricane Igor*. Government of Newfoundland and Labrador. http://www.stats.gov.nl.ca/maps/PDFs/HurricaneIgor.pdf.

Newkirk, R.T. 2001. "The Increasing Cost of Disasters in Developed Countries: A Challenge to Local Planning and Government." *Journal of Contingencies* and *Crisis Management* 9 (3): 159–70. http://dx.doi.org/10.1111/1468-5973.00165.

Ogrodnik, I. 2013. "By the Numbers: 2013 Alberta Floods." *Global News*, 26 June. http://globalnews.ca/news/673236/by-the-numbers-2013-alberta-floods/.

Oh, N., A. Okada, and L.K. Comfort 2014. "Building Collaborative Emergency Management Systems in Northeast Asia: A Comparative Analysis of the Roles of International Agencies." *Journal of Comparative Policy Analysis: Research and Practice* 16 (1): 94–111. https://doi.org/10.1080/13876988.2013.863639.

O'Leary, R., and L. Bingham, eds. 2009. *The Collaborative Public Manager: New Ideas for the Twenty-First Century*. Washington, DC: Georgetown University Press.

Olson, M. 1965. *The Logic of Collective Action*. Cambridge: Harvard University Press.

Olson, R.S., R.A. Olson, and V.T. Gawronski. 1998. "Night and Day: Mitigation Policymaking in Oakland, California. Before and After the Loma Prieta Disaster." *International Journal of Mass Emergencies and Disasters* 16 (2): 145–79. http://ijmed.org/articles/183/download/.

Olsson, E.-K. 2014. "Crisis Communication in Public Organisations: Dimensions of Crisis Communication Revisited." *Journal of Contingencies and Crisis Management* 22 (2): 113–25. http://dx.doi.org/10.1111/1468-5973.12047.

Owen, C., B. Brooks, C. Bearman, and S. Curnin. 2016. "Values and Complexities in Assessing Strategic-Level Emergency Management Effectiveness." *Journal of Contingencies and Crisis Management* 24 (3). http://dx.doi.org/10.1111/1468-5973.12115.

Parisien, M.-A., V.S. Peters, Y. Wang, J.M. Little, E.M. Bosch, and B.J. Stocks. 2006. "Spatial Patterns of Forest Fires in Canada, 1980–1999." *International Journal of Wildland Fire* 15 (3): 361–74. https://doi.org/10.1071/WF06009.

Parsons, B.M. 2012. "Making Connections: Performance Regimes and Extreme Events." *Public Administration Review* 73 (1): 63–73. http://dx.doi.org/10.2307/23355438.

Pasch, R.J., and T.B. Kimberlain. 2011. *Tropical Cyclone Report: Hurricane Igor (AL112010)*. National Hurricane Center. http://www.nhc.noaa.gov/data/tcr/AL112010_Igor.pdf.

Perry, R.W. 2007. "What Is a disaster?" In H. Rodriguez, E. Quarantelli, and R.R. Dynes, eds., *Handbook of Disaster Research*. New York: Springer.

Perry, R.W., and E.L. Quarantelli. 2005. *What Is a Disaster? New Answers to Old Questions*. Philadelphia: Xlibris.

Peters, E.J., and D.A. McEntire. 2010. *Emergency Management in Australia: An Innovative, Progressive and Committed Sector*. Emmitsburg, MD: Federal Emergency Management Agency.

Postmedia News. 2010. "Military Ends Hurricane Igor Relief Effort." *Global News*, 4 October. https://globalnews.ca/news/98664/military-ends-hurricane-igor-relief-effort/.

Prince, S.H. 1920. *Catastrophe and Social Change: Based upon a Sociological Study of the Halifax Disaster*. New York: Longmans, Green.

Public Safety Canada. n.d.. "About Public Safety Canada." Government of Canada. Accessed 8 August 2017. https://www.publicsafety.gc.ca/cnt/bt/index-en.aspx.

– 2013. "Harper Government Is Helping Manitoba Recover from the 2011 Flooding." News Release, 17 September. https://www.publicsafety.gc.ca/cnt/nws/nws-rlss/2013/20130917-eng.aspx.

– 2015. *Floods*. Government of Canada. https://www.publicsafety.gc.ca/cnt/mrgnc-mngmnt/ntrl-hzrds/fld-eng.aspx.

Qu, S., and J. Dumay. 2011. "The Qualitative Research Interview." *Qualitative Research in Accounting & Management* 8 (3): 238–64. http://dx.doi.org/10.1108/11766091111162070.

Raikes, J., and G. McBean. 2016. "Responsibility and Liability in Emergency Management to Natural Disasters: A Canadian Example." *International Journal of Disaster Risk Reduction* 16: 12–18. https://doi.org/10.1016/j.ijdrr.2016.01.004.

Redekop, B. 2011. "Province's Flood Predictions Off: Engineer." *Winnipeg Free Press*, 18 May. http://www.winnipegfreepress.com/breakingnews/Province-miscalculated-flooding-engineer-122151429.html.

Reid, L.D. 2015. *Op LENTUS in Review*. Canadian Army. http://www.army-armee.forces.gc.ca/en/news-publications/western-news-details-secondary-menu.page?doc=op-lentus-in-review/idw3e8gf [Link accessed 2018].

Rimstad, R., O. Njå, E.L. Rake, and G.S. Braut. 2014. "Incident Command and Information Flows in a Large-Scale Emergency Operation." *Journal of Contingencies and Crisis Management* 22 (1): 29–38. https://doi.org/10.1111/1468-5973.12033.

Ring, T.J. 2009. "Civil-Military Relations in Canada: A 'Cluster Theory' Explanation." MA thesis. Royal Military College of Canada.

Rodriguez, H., E. Quarantelli, and R.R. Dynes, eds. 2007. *Handbook of Disaster Research*. New York: Springer.

Rosow, I.L. 1955. "Conflict of authority in natural disaster." PhD diss. Harvard University.

Rural Municipality of Macdonald. 2011. "Meeting Minutes (May–July 2011)." http://www.rmofmacdonald.com/index.asp?ID=2&Sub_ID=144&Sub2_ID=0 [Link accessed 2018].

Rural Municipality of St. Francois Xavier. 2011. "Meeting Minutes (May–July 2011)." http://www.rm-stfrancois.mb.ca/main.

aspx?CategoryCode=BE0B3259-5572-4DEA-9D08-6072C1F49D90
&pageCode=773529F5-397B-4964-AA0A-85CD1F0641CC#.VunRrMfp6f5.

Saab, D.J., A. Tapia, C. Maitland, E. Maldonado, and L.-M. Ngamassi Tchouakeu. 2013. "Inter-organizational Coordination in the Wild: Trust Building and Collaboration among Field-Level ICT Workers in Humanitarian Relief Organizations." *Voluntas* 24 (1): 194–213. https://doi.org/10.1007/s11266-012-9285-x.

Saideman, S.M. 2016. *Adapting in the Dust*. Toronto: University of Toronto Press.

Salmon, P., N. Stanton, D. Jenkins, and G. Walker. 2011. "Coordination during Multi-agency Emergency Response: Issues and Solutions." *Disaster Prevention and Management* 20 (2): 140–58. https://doi.org/10.1108/09653561111126085.

Savage, E., M.D. Christian, S. Smith, and D. Pannell. 2015. "The Canadian Armed Forces Medical Response to Typhoon Haiyan." *Canadian Journal of Surgery* 58 (3): S146–52. https://doi.org/10.1503/cjs.013514.

Scanlon, J. 1982. "The Roller Coaster Story of Civil Defence Planning in Canada." *Emergency Planning Digest* 9 (2): 2–14.

– 1988. "Disaster's Little Known Pioneer: Canada's Samuel Henry Prince." *International Journal of Mass Emergencies and Disasters* 6 (3): 213–32. http://ijmed.org/articles/516/.

– 1995. "Federalism and Canadian Emergency Response: Control, Co-operation and Conflict." *Australian Journal of Emergency Management* 10 (1): 18–24. https://search.informit.org/doi/10.3316/informit.404893872653851.

– 1998. "Military Support to Civil Authorities: The Eastern Ontario Ice Storm." *Military Review* 4: 41–51.

– 2005. "Canadian Military Emergency Response: Highly Effective, but Rarely Part of the Plan." *Military Review* 85 (6), 74–9.

– 2007. "Unwelcome Irritant or Useful Ally? The Mass Media in Emergencies." In H. Rodriguez, E. Quarantelli, and R.R. Dynes, eds., *Handbook of Disaster Research*. New York: Springer.

– 2012. "Parallel Planning, Local Solutions: Four Newfoundland Airports Deal Diverted Flights after 9/11." *Journal of Contingencies and Crisis Management* 20 (3): 180–8. https://doi.org/10.1111/j.1468-5973.2012.00669.x.

Scanlon, J., E. Steele, and A. Hunsberger. 2012. "By Air, Land, and Sea: Canada Respond to Hurricane Katrina." *Canadian Military Journal* 12 (3): 54–62. http://www.journal.forces.gc.ca/vol12/no3/doc/PDFeng/Scanlon-Steele-Hunsberger%20Page5462.pdf.

Schmeltz, M.T., S.K. González, L. Fuentes, A. Kwan, A. Ortega-Williams, and L.P. Cowan. 2013. "Lessons from Hurricane Sandy: A Community Response in Brooklyn, New York." *Journal of Urban Health* 90 (5): 799–809. https://doi.org/10.1007/s11524-013-9832-9.

Schroeder, A., G.L. Wamsley, R. Ward. 2001. "The Evolution of Emergency Management in America: From a Painful Past to a Promising but Uncertain Future." In A. Farazmand, ed., *Handbook of Crisis and Emergency Management*. New York: Marcel Decker.

Schwartz, L., M. Hunt, L. Redwood-Campbell, and S. de Laat. 2014. "Ethics and Emergency Disaster Response: Normative Approaches and Training Needs for Humanitarian Health Care Providers." In D.P. O'Mathúna, B. Gordijn, and M. Clarke, eds., *Disaster Bioethics: Normative Issues When Nothing Is Normal*. Dordrecht: Springer Netherlands.

Seawright, J., and J. Gerring. 2008. "Case Selection Techniques in Case Study Research: A Menu of Qualitative and Quantitative Options." *Political Research Quarterly* 61 (2): 294–308. https://doi.org/10.1177/1065912907313077.

Segal, H. 2002. "The Canada-U.S. Defence Relationship: Nostalgia Ain't What It Used to Be." *Policy Options*, 1 April. http://policyoptions.irpp.org/magazines/continental-defence/the-canada-us-defence-relationship-nostalgia-aint-what-it-used-to-be/.

Seppänen, H., J. Mäkelä, P. Luokkala, and K. Virrantaus. 2013. "Developing Shared Situational Awareness for Emergency Management." *Safety Science* 55: 1–9. https://doi.org/10.1016/j.ssci.2012.12.009.

Shemella, P. 2006. "The Spectrum of Roles and Missions of the Armed Forces." In T.C. Bruneau and S.D. Tollefson, eds., *Who Guards the Guardians and How: Democratic Civil–Military Relations*. Austin: University of Texas Press.

Simpson, E. 2011. "From Inter-Dependence to Conflation: Security and Development in the Post-9/11 Era." *Canadian Journal of Development Studies* 28 (2): 263–75. https://doi.org/10.1080/02255189.2007.9669205.

Sjoberg, G. 1962. "Disasters and Social Change." In G. Baker and D. Chapman, eds., *Man and Society in Disaster*. New York: Basic Books.

Skertich, R.L., D.E.A. Johnson, and L.K. Comfort. 2012. "A Bad Time for Disaster: Economic Stress and Disaster Resilience." *Administration & Society* 45 (2): 145–66. https://doi.org/10.1177/0095399712451884.

Sloan, F.A., L.M. Chepke, and D.V. Davis. 2013. "Race, Gender, and Risk Perceptions of the Legal Consequences of Drinking and Driving." *Journal of Safety Research* 45: 117–25. https://doi.org/10.1016/j.jsr.2013.01.007.

Smith, S. 2017. "Army's Offer to Help with Kanesatake Flooding Revives Memories of Oka Crisis." *CBC News*, 9 May. http://www.cbc.ca/news/canada/montreal/army-s-offer-to-help-with-kanesatake-flooding-revives-memories-of-oka-crisis-1.4106827.

Smith, S., and S. Marandola. 2017. "Armed Forces to Stay in Quebec Even after Waters Recede, Philippe Couillard Says." *CBC News*, 11 May. http://www.cbc.ca/news/canada/montreal/quebec-flooding-rain-1.4109769.

Snyder, R. 2001. "Scaling Down: The Subnational Comparative Method." *Studies in Comparative International Development*, 36 (1): 93–110. http://dx.doi.org/10.1007/BF02687586.

Solomon, E. 2017. "Disaster Politics and the Floods of 2017." *Macleans*, 13 May. http://www.macleans.ca/news/canada/disaster-politics-and-the-floods-of-2017/.

Somers, S., and J.H. Svara. 2009. "Assessing and Managing Environmental Risk: Connecting Local Government Management with Emergency Management." *Public Administration Review* 69 (2): 181–93. https://doi.org/10.1111/j.1540-6210.2008.01963.x.

Stake, R.E. 2010. *Qualitative Research: Studying How Things Work*. New York: Guilford Press.

Stallings, R.A. 2007. "Methodological Issues." In H. Rodriguez, E. Quarantelli, and R.R. Dynes, eds., *Handbook of Disaster Research*. New York: Springer.

Stevenson, J.R., Y. Chang-Richards, D. Conradson, S. Wilkinson, J. Vargo, E. Seville, and D. Brunsdon. 2014. "Organizational Networks and Recovery following the Canterbury Earthquakes." *Earthquake Spectra* 30 (1): 555–75. http://dx.doi.org/10.1193/022013EQS041MR.

Stocks, B.J. et al. 2002. "Large Forest Fires in Canada, 1959–1997." *Journal of Geophysical Research* 108 (D1): 8149. https://doi.org/10.1029/2001JD000484.

Stoltenberg, J. 2015. "The Secretary General's Annual Report." *NATO*, 30 January. http://www.nato.int/cps/en/natohq/opinions_116854.htm.

Stoney, C., and K.A.H. Graham. 2009. "Federal-Municipal Relations in Canada: The Changing Organizational Landscape." *Canadian Public Administration* 52 (3): 371–94. https://doi.org/10.1111/j.1754-7121.2009.00088.x.

Stunden Bower, S. 2011. "The Assiniboine River Flood of 2011: Without Precedent?" *NiCHE*, 15 May. http://niche-canada.org/2011/05/15/the-assiniboine-river-flood-of-2011-without-precedent/.

Susanto, N., S. Nugroho, and E. Rizkiyah. 2018. "Evaluating Risk Perception Based on Gender Differences for Mountaineering Activity." *E3S Web of Conferences* 31 (3): 09028. https://doi.org/10.1051/e3sconf/20183109028.

Tatham, P., R. Oloruntoba, and K. Spens. 2012. "Cyclone Preparedness and Response: An Analysis of Lessons Identified Using an Adapted Military Planning Framework." *Disasters* 36 (1): 54–82. https://doi.org/10.1111/j.1467-7717.2011.01249.x.

Tatham, P., and S.B. Rietjens. 2016. "Integrated Disaster Relief Logistics: A Stepping Stone towards Viable Civil–Military Networks?" *Disasters* 40 (1): 7–25. https://doi.org/10.1111/disa.12131.

Tatham, P., K. Spens, and G. Kovács. 2016. "The Humanitarian Common Logistic Operating Picture: A Solution to the Inter-agency Coordination Challenge." *Disasters* 41 (1): 77–100. https://doi.org/10.1111/disa.12193.

Thompson, W.C. 2010. "Success in Kashmir: A Positive Trend in Civil–Military Integration during Humanitarian Assistance Operations." *Disasters* 34 (1): 1–15. https://doi.org/10.1111/j.1467-7717.2009.01111.x.

van Laere, J. 2013. "Wandering through Crisis and Everyday Organizing: Revealing the Subjective Nature of Interpretive, Temporal and Organizational Boundaries." *Journal of Contingencies and Crisis Management* 21 (1): 17–25. https://doi.org/10.1111/1468-5973.12012.

Vidal, R. 2015. "Managing Uncertainty: The Engineer, the Craftsman and the Gardener." *Journal of Contingencies and Crisis Management* 23 (2): 106–16. https://doi.org/10.1111/1468-5973.12081.

Vigoda, E., and E. Gilboa. 2002. "The Quest for Collaboration: Toward a Comprehensive Strategy for Public Administration." In E. Vigoda, ed., *Public Administration: An Interdisciplinary Critical Analysis*. New York: Marcel Dekker.

Voorhees, W.R. 2008. "New Yorkers Respond to World Trade Centers Attack: An Anatomy of an Emergent Volunteer Response." *Journal of Contingencies and Crisis Management* 16 (1): 3–13. https://doi.org/10.1111/j.1468-5973.2008.00530.x.

Wachtendorf, T., and J. Kendra. (2005). *A Typology of Organizational Response to Disasters*. Presentation to the American Sociological Association, Philadelphia, PA, 14 August.

Walker, P. 1992. "Foreign Military Resources for Disaster Relief: An NGO Perspective." *Disasters* 16 (2): 152–9. https://doi.org/10.1111/j.1467-7717.1992.tb00389.x.

Warner, R. 2013. "Resilience or Relief: Canada's Response to Global Disasters." *Canadian Foreign Policy Journal* 19 (2): 223–35. https://doi.org/10.1080/11926422.2013.773541.

Waugh, W., Jr., and K. Tierney, eds. 2007. *Emergency Management: Principles and Practice for Local Government*, 2nd ed. Washington, DC: International City Managers Association.

Wells, K.B. et al. 2013. "Applying Community Engagement to Disaster Planning: Developing the Vision and Design for the Los Angeles County Community Disaster Resilience Initiative." *American Journal of Public Health* 103 (7): 1172–80. https://doi.org/10.2105/ajph.2013.301407.

Wickramasinghe, N., R.K. Bali, and R.N.G. Naguib. 2006. "Application of Knowledge Management and the Intelligence Continuum for Medical Emergencies and Disaster Scenarios." *International Conference of the IEEE Engineering in Medicine and Biology Society* 2006: 5149–52. https://doi.org/10.1109/iembs.2006.260825.

*The Wildfire Act, Statues of Saskatchewan* 2014, c W-13.01. https://www.saskpublicsafety.ca/-/media/project/spsa/documents/reports/the-wildfire-act.pdf.

Williams, B.H. 1956. *Emergency Management of the National Economy*. Washington, DC: Industrial College of the Armed Forces.

Willing, J. 2017. "Between Gas Leak and Flooding, City's Emergency Management Team Going 'Flat Out.'" *Ottawa Sun*, 19 May. http://www.ottawasun.com/2017/05/19/between-gas-leak-and-flooding-citys-emergency-management-team-going-flat-out.

Zhang, H., X. Zhang, L. Comfort, and M. Chen. 2016. "The Emergence of an Adaptive Response Network: The April 20, 2013 Lushan, China Earthquake." *Safety Science* 90: 14–23. https://doi.org/10.1016/j.ssci.2015.11.012.

Zurcher, L.A. 1968. "Social-Psychological Functions of Ephemeral Roles: A Disaster Work Crew." *Human Organization* 27 (4): 281–97. https://doi.org/10.17730/humo.27.4.m055343261781736.

# Index

Page references to charts and tables are in bold.

Aboriginal Affairs and Northern Development Canada (AANDC), 57
after action reviews (AAR), 103
Aid to Civil Authority, 75, 203n19, 205n37, 207n10
Aid to Law Enforcement, 75
Alberta: emergency management, 59, 89; Provincial Operations Centre (POC), 59
Alberta multi-river flood, 53; archival sources, 182; assets deployed, 60, 78; CAF deployment, 59–60, 68–9, 70–1, 120, 142, 203n19; civil–military relations, 112–13, 119; decision-making processes, 96; duration, 58; empire-building context, 116; evacuation orders, 58–9, 60, 71, 120; federal response, 59; hazard, 58; historical context, 129; impact, 58, 128; location, 58; municipal and provincial response, 59; Request for Assistance, 128, 140; sandbagging operation, 118–19
all-hazards approach, 10, 143, 159, 160, 161, 162, 189

American Federal Response Plan, 11
Aurora aircrafts, 78, 79
Australia: disaster response system, 11–12, 145, 212n5; emergency management framework, 11–12; natural hazards, 11, 196n1

barriers to interorganizational collaboration: assessment of, 6, 7, 110; benign incapacity, 119–25; case studies, **186–7**; chain of command conceptual difference as, 130, 131; components of, 7–8, **49, 133–4,** 135, **137,** 168, **168,** 170–1; concept of, **4**; empire-building, 110–17; front-line interoperability, 149; indicators for the components, 138; manipulation and distrust, 117–19; RFA process and, 74, 204n22
battle rhythm, 81, 127
benign incapacity: definition, 119–20, 167; federal institutional constraints and, 124–5; indicators for, 110, 171; institutional context, 144–5; military as antidote to, 120–2, 135; mitigation of, 122;

occurrences of, 48, 124; of RCAF, 145; resource constraints and, 122–4
Boin, A., 189
Boston Marathon bombing, 189
Brandon, MB: CAF deployment in, 71; civil–military relations, 112; displaced population, 56; emergency management, 66, 96, 112–13, 122, 202n10; floods, 15–16
bureau: definition of, 199n20; economic theories of, 199n22

Calgary, AB: civil–military relations, 112; emergency management programs, 66–7, 112–13, 122; Emergency Operation Centre, 15; evacuation order, 59
Canada: Afghanistan mission, 31; defence priorities, 30; emergency management capacity of territories, 194n9; federal–municipal divide, 12; foreign policy, 33; military command structure, 32; NATO defence spending, 30
Canada Border Services Agency (CBSA), 20
Canada Revenue Agency (CRA), 55
Canadian Armed Forces (CAF): academic analysis of, 36; assets, 77–80, 204n24; branches, 29; bureaucracy, 36; civilian benign incapacity and, 120–1; collegial networking, 95; community impact, 106; in comparative perspective, 5; decision-making processes, 154; democratic accountability of, 41, 204n23; disaster response role of, 6, 24, 28, 35, 36, 49, 82, 97, 99, 111–12, 153, 171–2, 196n27; effectiveness of, 6, 30–1, 35, 101, 203n19, 208nn16, 17; in emergency management, 4–5, 27–8, 35, 37; end state vision, 91–2, 100–1; establishment of, 196n24; ethos of, 115–16; exercises, 81, 156; federal government and, 20–2; flexibility of, 74–5; in flood response, 70–1; funding of, 121, 146, 206n40; governance of, 22; historical traditions, 29; law enforcement and, 207n9; local contractors and, 71; mandate, 5, 28–9, 88; media relations, 102; municipal relationships, 95–6, 105; non-combat mission, 29, 33; openness of, 112; orders format, 82, 206n39; organizational structure, 28; overseas operations, 29, 31, 32; public perception of, 106, 152, 153; radio systems, 205n37; regular and reserve force, 29; reputation of, 102, 115; resource allocation, 71, 75, 82–3, 152; response capacity, 17, 212n6; security support, 33, 72–3, 207n10; self-sufficiency of, 83; soldiers' qualifications, 28; tactical actors, 104; technological support, 80–3; tools and skills of, 28; training system of, 121–2, 195n14; unit-level empires, 116–17; in wildfire management, 65–6
*Canadian Human Rights Act*, 76
Canadian Joint Operations Command (CJOC), 29, 82
Canadian Security Intelligence Service (CSIS), 20, 157
Canadian Special Operations Forces Command, 29
CF-188 Hornets, 196n25
Chinook helicopters, 78
civil defence, 13
civilian emergency managers, 88, 90–1, 92, 101, 102

civilian–military relations: collaboration in, 41–2; complexity of, 40–1; conflict resolution, 83, 86; data sharing, 107; decision-making process, 207n11; in democratic context, 41; division of labour, 41; effectiveness of, 147; empire-building and, 112; gender and ethnic factors of, 163; normative assessment of, 8; openness of, 112; problems of, 94, 107, 109; satisfaction in, 109; shared responsibility, 42, 43, 46, 47, 154; studies of, 36, 41, 162, 163–4; trust in, 107, 109

civilian officials: disaster response approach, 89; emergency management, 88; vision of end state, 88

climate change, 13

collaboration: barriers to, 48; studies of, 38

collaborative framework, **49,** 147, 148–9, 166

collaborative governance regime, 46–7

collegial networking, 94–5

common operating picture (COP), 80–1, 205n36

Communications Security Establishment, 22

competition: among public sector organizations, 150; benefits and costs of, 151; interorganizational, 37–8; of intragovernmental bureaus, 113, 151

complex events, 26

conceptual difference: along military–civilian lines, 126–7, 130–1, 132; within CAF, 128–30; ICS vs. chain of command, 131–2; indicators for, 110, 125–6, 171; "pure," 126, 130; strategy vs. tactics, 127; wildfire management, 130–1

conflict resolution, 40, 46, 83, 86, 169, 213n1

coordination, 38, 43–4, 139, 169–70; vs. integration, 208n14

Department of National Defence (DND), 22, 27, 55, 95

disaster: academic research, 197n2; definition of, 24, 25, 50, 175, 195n18; industrial, 162, 163; responsibility for management of, 3; security-related, 162, 163

disaster events, 26, 47, 51, 53–5, 173

disaster response: academic research of, 6, 22–3, 50–1, 172, 173, 197n6; after action review, 103; air support challenges, 78–9; case studies, 51–3; civilian–military relations in, 162–3, 171–2; Cold War and, 10; collegial network and, 94; consultation prior to action, 97; coordination, 23–4, 157, 188–9; cross-country comparison, 11–12, 145–6; decentralized system of, 157–8; decision-making process, 96, 99, 100, 154, 157; development of, 10, 27, 153; effectiveness of, 47, 151, 155, 189, 192, 198n14; federal capacity for, 5, 8–9, 145–7; in First Nation communities, 21; formal rules in, 144; front-line actions, 148; human resources and, 76–7; institutional role in, 144; "internal" and "external" dynamics, 139–40; interorganizational competition and, 151; military role in, 6, 22, 24, 27, 28, 35, 36, 49, 82, 97, 99, 111–12, 153, 171–2, 196nn20, 27;

non-government sectors and, 198n17; operational problem, 24, 189; organizational behaviour during, 150; planning process, 100; policy recommendations on, 9, 154–60, 161; politics and, 194nn5, 7, 195n16, 212n4; situational understanding, 103–5, 106; strategic and tactical actors, 103–4, 105; timelines, 78; visions of, 92–3
disaster response organizations: collaboration of, 39, 189; decision-making processes, 99; lack of integrated activities, 98–9; purpose of, 98; relevant actors of, **167**; solidarity between, 97, 98
disaster response study: hypotheses of, 213n2; interviews, 172–5, 177, **178–81**, 213nn3, 6; participants, 175–7, 213n4; snowball recruitment technique, 176, 214n8; theoretical framework, 177
distrust, 48, 110, 117, 118–19, 171, 211n13
domestic operations doctrine, 31–2, 42–4, 47, 79, 118, 206n3
Donahue, A.K., 189
Drabek, T.E., 189

economic theory of the bureau, 47, 48, 150, 199n22
effectiveness: definitions of, 197n7. *See also* measures of effectiveness
emergencies, 26
emergency management (EM): all-hazards approach, 10, 159, 189; case studies, 49–53; cities and, 11; climate change and, 13; decentralization of, 157–8; definition of, 26; democratic accountability and, 75; at federal level, 3, 12, 13, 19; federal–municipal relations and, 3, 12; function of, 202n9; historical development of, 5; interorganizational collaboration and, 43, 46, 47, 93, 190; interorganizational competition and, 38–9, 151; legislation on, 12, 26; military contribution to, 4–5, 26–7, 35, 36–7; at municipal level, 3, 193n1(Ch1); national security and, 12; networks of, 155–6; nuclear attack-focused, 10; phases of, 155, 193n2(Int); policy implementation, 45, 49–50, 125; at provincial level, 3, 11, 124–5; public administration of, 5, 11, 19, 45; scholarly literature on, 6, 166, 188–9
*Emergency Management Act*, 12, 25, 26, 194n6, 195n14
emergency management organizations (EMOs): collaboration among, 189; creation of, 5; criticism of, 190–1; disaster response and, 23, 157–8; empire-building problems, 113, 211n17; funding of, 195n12; hierarchy of, 13–14, 93, 190–1; level of independence of, 194n8; measures of success, 208nn16–18; name recognition, 194n8; provincial, 18–19, 113, 190
Emergency Operations Centres (EOCs): Canadian Armed Forces in, 4; capacity of, 16–17; influence on decision making, 195n13; interorganizational collaboration in, 39; levels of, 156; liaison officers in, 64–6, 90, 111, 140; mandate, 14, 202n5; municipal, 15–16, 195n13; policy recommendation, 155–6; provincial, 5, 195n13
empire-building: as barrier to collaboration, 48; bureaucratic competition and, 112–13; within

civilian levels of government, 114, 115, 150, 211n17; in civil–military relations, 111–12, 113, 150; in context of trust, 211n15; indicators for, 110, 114, 170, 210n8; individual priorities and, 114; information sharing and, 113–14; in the military, lack of, 113, 115–17; organizational prestige and, 111; potential reasons for, 210n5; unit-level, 115, 116
end state: civilian vs. military vision of, 87–8, 89, 90–2, 98, 100–1, 151; conceptual differences regarding, 126–7, 136, 152
evacuation notices: vs. evacuation orders, 200n36

Federal Emergency Management Agency (FEMA), 3, 20, 145–6
federal government: in disaster response, 145–7, 160; empire-building in, 114; municipal relations, 11; responsibilities of, 19
firefighters, 90, 160, 209n21
First Nations: disaster response, 21, 58; evacuation orders, 203n17; federal government and, 21; flood management, 59, 200n37; military and, 33; Saskatchewan wildfires and, 60
flexibility, 70, **84**, 169, 203n18
flood management, 15–16, 65, 89–90
frigates, 80
Fritz, C., 23, 188
front-line disaster response, 148, 149, 163

government agencies: competition between, 37–8; conflict between civilian levels of, 206n41
Government Operations Centre (GOC), 21, 114, 145, 210n6

Griffon helicopters, 78
Guskin, S.L., 23, 188

Haiti earthquake, 189
Halifax harbour explosion, 22, 27, 51, 163, 199n24
hard assets (machines): aircrafts, 78–9, 123, 203n20, 205nn30, 31, 32, 34, 213n7; civilian demand for, 77–8, 86, 159; effects of, 124; frigates, 80; helicopters, 78; military vehicles, 204n29; perception of capabilities of, 123–4; psychosocial impact of deployment of, 80, 203n20
Harper, Stephen, 67, 202n13
hazards, 8, 11
Health Canada, 207n8
health emergency management (HEM), 113, 114, 210n7, 211n17
Hercules aircrafts, 78–9, 123, 203n20, 205n31
Her Majesty's Canadian Ships (HMCS), 29
Herspring, D.R., 35
High River flood, 58, 59, 70–1, 72
human resources, 75–7, 78, 204n25
Hunsberger, A., 205n35
Huntington, Samuel, 41
Hurricane Igor, 53; air support, 78; archival sources, 182; bridge-building, 118; CAF deployment, 55, 68, 71, 120; civil–military collaboration, 119; conflict resolution, 83, 86; decision-making processes, 96; duration of, 54; effectiveness of command during, 205n36; empire-building context, 116; federal response to, 55; front-line interpretation of, 130, 148; hazard of, 54, 143; historical context, 129; impact of, 54, 55, 65, 200nn29, 33; information sharing

during, 64; municipal response to, 54–5; political issues and, 67–8, 194n7; provincial response to, 55; Request for Assistance, 68, 128, 140; reservist officers during, 67, 68–9; support from EMOs, 113
Hurricane Katrina, 189
Hydrologic Forecast Centre (HFC), 57

Incident Command System (ICS), 105, 131, 148, **165**, 190–1
Indigenous Services Canada, 21
indoctrination, 210n12
informal networks, 8, 140–2, 160
information sharing: in CAF's strategy, 209n2; empire-building and, 113–14; hazard type and, 65–6; indicators for, 62–3, **84**, 168–9; liaison officers' role in, 63–5; manipulation and, 117; municipal levels of, 66
institutional constraints, 70, 124–5, 126, 204n22
integration: characteristics of, 97; civilian–military, 139; vs. coordination, 208n14; definition of, 44, 212n3; indicators of, 147–8, 149, 170; limits of, 147–9
interorganizational collaboration: actors of, **167**; assessment of, 136–40; challenges of, 48, 188–9; components of, 39–40; emergency management and, 43, 46, 47, 93, 190; informal networks and, 141; institutional context, 144; "internal" and "external" dynamics of, 139–40; limits of, 149; literature on, 162, 166, **166**, 188–9; between military and civilian levels, 61, 152; non-hierarchical nature of, 93–4; study of, **166**, 171–3
Islamic State of Iraq and the Levant (ISIL), 30

Johnston, David, 164
Joint Task Force Atlantic (JTFA), 63, 68, 82, 94, 121, 128, 129
Joint Task Forces (JTFs): disaster response, 5, 63, 157–8; domestic operations exercises, 94; food provision responsibility, 119; liaison officers of, 63; orders format, 82; regional, 29–30, 129; reservist officers of, 68; RFA process and, 52, 128; training system, 121
Joint Task Force West (JTFW), 63, 68, 94, 119, 121, 128, 129

Kuipers, S., 189

Lac-Mégantic train derailment, 199n26
leaders: military vs. civilian, 88
lessons learned processes, 156
liaison officers (LOs): in CAF doctrine, 63; in combat context, 63; in domestic context, 63–4; at municipal level, 66–7; in provincial EOCs, 64–6, 90, 111, 140; responsibilities of, 63–5, 100, 140, 207n6
Light Armoured Vehicles (LAVs), 71, 77–8, 95, 124

manipulation, 48, 110, 117, 170
Manitoba Assiniboine River flood, 53; air support, 78, 79; archival sources, 182; CAF deployment, 57–8, 71, 73–4, 120; civil–military collaboration, 112–13, 119; conceptual difference in approach to, 129–30; conflict resolution, 86; decision-making processes, 96; duration, 56; empire-building context, 116; federal response to,

57; front-line interpretation of, 130, 148; hazard of, 56, 65, 128; impact of, 56, 65; municipal response to, 56–7; provincial response to, 57; Request for Assistance, 69, 128; reservist officers, 69; sandbagging operation, 71, 118–19, 120
Manitoba Emergency Coordination Centre (MECC), 57
McLuckie, B.F., 23
measures of effectiveness, 100–1, 102, 103
Medium Support Vehicle System (MSVS), 95, 204n29
military officers, 90, 91, 92, 100–1
military retirees, 214n10
mission command doctrine, 130, 158
Moore, H.E., 195n16
mounted infantry platoons, 71, 204n28
municipal government: decision-making capacity of, 105; emergency management capabilities, 73–4, 159–60; funding of, 159–60; jurisdiction of, 13, 15; public policy and, 14–15; resource constraints, 122–3

natural disasters, 3, 18, 162, 193n2(Ch1), 196n1
networking processes, 94
New Brunswick ice storm, 129
Newfoundland and Labrador: federal politics and, 203n13. *See also* Hurricane Igor
new public governance (NPG): actors of, **167**; definition of, 45–6; goal of, 191, 192; management practices of, 191–2, 215n16; normative goals of, 215n17; study of, 166
9/11 terrorist attacks, 12, 151, 199n21
non-evacuees, 207n9, 211n19

non-manipulative influence, 39, 46, 70, 74, **84**, 138, 169
non-routine emergencies, 25, 52
North American Aerospace Defense Command, 29
North Atlantic Treaty Organization (NATO), 5, 30–1

October Crisis, 33
O-groups, 64, 81–2, 99–100, 202n6
Oka Crisis, 33
Ontario: CAF deployment, 17, 18; ice storm of 1998, 22
Operation IMPACT, 31, 205n33
Operation LAMA, 55
Operation LENTUS, 31, 59–60, 61
Operation LIMPID, 32
Operation LUSTRE, 57–8
Operation NANOOK, 32
Operation NEVUS, 32
Operation PALACHI, 31
Operation REASSURANCE, 31
organizational anti-collaboration, 47, 48

physical resources. *See* hard assets (machines)
preparedness: vs. emergency management, 19
presence of interorganizational collaboration: analysis of, 6–7; assessment of, 75; case studies, **184–5**; civilian–military relationship in, **137**; collective conflict resolution, 83, 86; components of, 7, **49**, 62, **84–5**, 127, 167, **167**, 168–9, **184–5**; concept of, 4; disaster response, **137**; indicators for the components, 7, 62, 70, 137–8; information sharing, 62–70; non-manipulative influence and flexibility, 70–5; support, 75–83

Prince, Samuel Henry, 51
provinces: "co-sovereign" status of, 17; disaster management, 17, 18
public administration, 44–5, 47
Public Health Agency of Canada (PHAC), 113, 114, 124–5, 145, 207n8
Public Safety Canada: collegial networking, 95; disaster response capacity, 125, 145; emergency management function of, 20, 124–5, 193n1(Int); information sharing and, 114; military and, 20–1; responsibilities of, 19, 20
public safety partner agencies, 195n11
public sector behaviour theories, 150–1

quality of interorganizational collaboration: assessment of, 6; case studies, **185–6**; civilian–military relationship, **137**; collegial networks, 92–7; components of, 7, **49, 108, 168,** 169–70, **185–6**; concept of, **4, 108,** 109; coordination, 87–97; end state visions, 87–92; indicators for the components, 87, 138; integration, 97–107, 127; measures of effectiveness, 100–3; satisfaction, 106; solidarity and separation, 98–100; strategic and tactical actors, 103–6; trust, 106–7
Quebec: floods, 18, 129; hazard management, 11

Rayner, J.F., 23, 188
Red Cross, 21, 24, 38, 198n17
Redford, Alison, 69
Red River Floodway, 209n3
Regional Municipality of Ottawa-Carleton (RMOC), 16

Request for Assistance (RFA): civilian vs. military understanding of, 135, 151; conceptual differences regarding, 126–7, 136, 151; confirmation process, 128–9; constraints of, 91–2, 111, 112; general tasks of, 73–4; guidelines for CAF actions, 70, 72–3, 195n14, 211n14; as institutional constraint, 70, 72–5, 86, 126, 204n22, 207n10; municipal leadership and, 73; official initiation of, 52, 72, 203n15; parameters of, 75; timing, 17–18, 68
reservist officers: informal networks of, 67–8, 69–70; liaison capacity of, 68, 69–70; status of, 201n3
Roblin, Duff, 210n3
routine events, 26
Royal Canadian Air Force (RCAF): benign incapacity of, 8, 145; civilian requests for resources of, 78–9, 123–4, 205n34; disaster response capacity of, 123, 203n17; NATO deployments, 79; resource constraints, 123, 152, 159; responsibilities of, 29, 79
Royal Canadian Mounted Police (RCMP), 20, 157, 204n26
Royal Canadian Navy (RCN): assets, 80, 205n35; concentration of resources of, 78; disaster response capacity of, 203n20; international training mission, 205n37; responsibilities of, 29
rules of engagement (ROEs), 72, 203n21
Russia, 30, 31

Salvation Army, 24, 38, 197n7, 198n17
Saskatchewan wildfires, 53; archival sources, 183; assets deployed, 61,

Index 249

79; barriers to interorganizational collaboration, 135; civilian–military collaboration, 89, 90, 119; conceptual difference, 131–2; decision-making processes, 96; duration of, 60; evacuation orders, 60; federal response, 61; fighting techniques, 142–3; First Nations and, 60; frontline non-interoperability, 148; hazard, 60, 89; impact of, 60; location of, 60; military involvement in fighting, 61, 65, 71, 79; municipal response, 60–1; provincial response, 61; reservist officers and, 69
Saskatchewan Wildland Fire Management branch, 65
satisfaction, 46, 106, 132, 170
Scanlon, Joe, 197n3, 205n35, 213n6
Schroeder, A., 23
Sea King helicopters, 78
security support, 30, 72–3
selflessness, 170
shared responsibility, 42, 43, 166, **167,** 198n13
situation reports (SITREPs), 81–2
Slave Lake fire, 59
snowball recruitment technique, 176, 214n8
soft assets, 80–3
Stelle, E., 205n35
strategy vs. tactics, 198n12
support, **85,** 139, 169

Synthetic-aperture radar (SAR) system, 205n34

't Hart, P., 189
Toronto Emergency Operation Centre, 16
transcend routine emergencies, 26
trust, 48, 170, 211n13
Tuohy, R.V., 189

United States: disaster response systems, 11, 212n5; emergency management, 11; natural hazards, 196n1
unity of effort, 97, 98
University of Colorado Boulder's Natural Hazard Center, 197n2
University of Delaware's Disaster Research Center (DRC), 197n2

Wachtendorf, Tricia, 213n6
Wamsley, G.L., 23, 24
Ward, R., 23, 24
"whole of government" concept, 57, 59, 61, 201n42
wildfires: all-hazards approach to, 143, 160; cost of, 201n44; evacuation orders, 203n17; fighting techniques, 142; hazard-specific effects of, 142–4, 160; management, 65–6, 130–1, 143, 160
Williams, Danny, 67–8, 194n7
Winnipeg floods, 15, 56, 209n3

www.ingramcontent.com/pod-product-compliance
Lightning Source LLC
Chambersburg PA
CBHW020250030426
42336CB00010B/696